Veterinary Ethics in Practice

Veterinary Ethics in Practice

James Yeates

CABI is a trading name of CAB International

CABI
Nosworthy Way
Wallingford
Oxfordshire OX10 8DE
UK

CABI
WeWork
One Lincoln St
24th Floor
Boston, MA 02111
USA

Tel: +44 (0)1491 832111
Fax: +44 (0)1491 833508
E-mail: info@cabi.org
Website: www.cabi.org

Tel: +1 (617)682-9015
E-mail: cabi-nao@cabi.org

A catalogue record for this book is available from the British Library, London, UK.

References to Internet websites (URLs) were accurate at the time of writing.

Library of Congress Cataloging-in-Publication Data

Names: Yeates, James, 1980- author.
Title: Veterinary ethics in practice / by James W. Yeates.
Description: Wallingford, Oxfordshire ; Boston, MA : CABI, [2021] | Includes
   bibliographical references and index. | Summary: "Concise, readable and
   accessible, Veterinary Ethics in Practice gives non-specialist veterinary
   professionals an introduction to ethics. It helps readers to think about, and
   discuss, ethical dilemmas and viewpoints faced by practitioners in their daily
   practice"-- Provided by publisher.
Identifiers: LCCN 2020031938 (print) | LCCN 2020031939 (ebook) | ISBN
   9781789247206 (paperback) | ISBN 9781789247213 (ebook) | ISBN
   9781789247220 (epub)
Subjects: LCSH: Veterinarians--Professional ethics.
Classification: LCC SF756.39 .Y43 2021 (print) | LCC SF756.39 (ebook) | DDC
   174.2/9--dc23
LC record available at https://lccn.loc.gov/2020031938
LC ebook record available at https://lccn.loc.gov/2020031939

ISBN-13: 9781789247206 (paperback)
         9781789247213 (ePDF)
         9781789247220 (ePub)

Commissioning Editor: Caroline Makepeace
Editorial Assistant: Ali Thompson
Production Editor: James Bishop

Typeset by SPI, Pondicherry, India
Printed and bound in the USA by Integrated Books International, Dulles, Virginia

Contents

Preface

This book is to help veterinary professionals develop the skills to deal with real issues, and engage in ethical reasoning and discussions with confidence. It is the result of many years of practising, policy-making, researching and wrestling with these issues. I make no claim to have 'solved' all of veterinary ethics – or to be better, morally, than anyone else. Indeed, the more I have reflected and learnt, the more I have realized my own failures.

This book therefore does not tell everyone what to do. Instead it presents different possible ways of thinking (and ways of thinking about thinking), with some of their key implications and challenges. You should disagree with many of the ideas presented – and then reflect on why you disagree (and what changes would make you agree). At the same time, please be open minded to changing your views several times as you read and reflect (sometimes back to where we started, but with greater confidence and clarity).

This book is structured around the kind of deliberations we might follow in practice. The first chapter introduces ethics. This is followed by Part A (Chapters 2–4), which considers various general topics. Part B (Chapters 5–9) considers key practical ethical skills. Part C (Chapters 10–13) applies this to practice, for various areas of veterinary work. Each chapter also highlights particular ideas and errors to consider or avoid. Except for Part B, each chapter has 'Reflections' or 'Applications'. I have kept these short and bite-size for busy people. So please spend time reflecting on and discussing the ideas, and formulating your own views, and applying them to *your* past, present and imaginary cases.

As a practical book, it avoids using technical language, delineating academic theories and parroting contemporary debates in ethics, medical ethics, metaethics and jurisprudence that, while interesting, are not particularly useful for us in our veterinary work. Similarly, I have avoided delineating which famous philosophers said what, which makes it particularly remiss in citations, and (even worse) seeming ungrateful to all the serious scholarship, but many are given in Further Reading, and I hope this taster helps you discover them.

The Author

Dr James Yeates studied veterinary medicine and medical ethics before working in clinical practice for ten years, while undertaking a PhD in veterinary ethics and a clinical training scholarship. He was previously Honorary Secretary of the Society of Practising Veterinary Surgeons, Chair of the British Veterinary Association's Ethics and Welfare Group, Chief Veterinary Officer of the Royal Society for the Prevention of Cruelty to Animals and a Royal College of Veterinary Surgeons Registered Specialist. He is currently Chief Executive of Cats Protection, a Diplomate of the European College of Animal Welfare and Behavioural Medicine, and a Fellow of the Royal College of Veterinary Surgeons.

Acknowledgements

Enormous thanks go to everyone who has helped: David Morton for inspiring the original idea; Madeline Campbell, Siobhan Mullan, Mandy Nevel, Ian Glover, Kit Sturgess and Sean Wensley for improving the document itself, and to Caroline Makepeace, Ali Thompson, and James Bishop at CABI for such great editorial support and copyeditor, Val Porter, for making it so much better.

Introduction: What Veterinary Ethics Is (Not)

> As veterinary professionals, in a variety of roles, our work presents us with a series of difficult moral challenges – sometimes too frequently for comfort. In many of our cases, we have to decide what we should do next. However, we can sometimes find it difficult to understand the situation, to process the emotions involved, to work out what to do, or to do it.
>
> If we recall challenging cases we have seen: what made them hard? How did we respond? How might ethics have helped us? Did we use ethical skills or just facts?

We face ethical challenges just as our patients face health challenges, as our environments and infective ideas from others interact with our internal 'ethical physiology'. We respond to ethical challenges through our behaviour, motivated by instincts and learning, although sometimes the right behaviour is not obvious or easy to explain or justify (even to ourselves). We also respond internally by making 'homeostatic' responses so that our behaviours fit with our moral beliefs, but sometimes we make 'allostatic' changes, fundamentally altering our views so that we respond differently the next time we are presented with similar stimuli. Ethics can help us ensure we make these internal changes in ways that improve our consistency, resilience and ability to cope with future challenges.

Ethics (and morality) is practical because it relates our reasoning to our behaviour (as opposed to speculative or theoretical reasoning) and because it is directly applicable to what we do. In practice, we have to decide what we should do next (Box 1.1). So ethics is part of veterinary work. Indeed, our ethical concerns are what give purpose and legitimacy to our work. Ultimately, ethics is the *ability* to decide well. Veterinary professionals are often experienced in making difficult, high-pressure decisions, but we can develop our ethical skills by reflection, discussion and education (as we do for other skills), in helping us act, influence and feel better (or at least less bad sometimes).

Box 1.1. We have to decide what we should do next

The phrase, 'We have to decide what we should do next' highlights several important aspects to veterinary ethics.

- **We** – We ultimately decide for ourselves, while listening to others, openly but critically.
- **Have to** – We cannot (as owners may) duck responsibility by 'letting nature take its course' or letting others decide for us. We have responsibilities to make decisions (and for the consequences if we do not). Indeed, when we avoid making a decision, we are responsible for the decision not to make it (while giving up the chance to affect the outcomes).
- **Decide** – We choose our actions, actively and consciously. Some ethical choices feel easy or obvious to experienced practitioners. However, more difficult, novel, complex or finely balanced decisions may require more explicit reasoning.
- **What** – Ethics is about concrete options in real situations: choosing behaviours (e.g. keeping promises and not stealing), characteristics (e.g. compassionate and honest) or outcomes (e.g. healthier patients).
- **We** – I can only ultimately control what I think and do. We can advise others, but cannot make everyone perfect or solve every problem.
- **Should** – We need not only descriptive facts, but also morally motivational reasons to act.
- **Do** (and not do) – Ethics is about action. Theory and even reflection are only helpful when applied to real cases.
- **Next** – We cannot know the future; all we can know is the right thing to do now. We can use the past to self-improve constructively, but not to self-chastise destructively (or to self-justify).

Some of us may find ethics uncomfortable. We might link it to scary legal or regulatory processes (note that this book is not a source of legal advice), or dislike uncertainty, disagreement, or questioning ourselves and previous behaviour. We might have seen ethical methods being misused in ways that seem unconvincing, unhelpful, sanctimonious or over-sentimental. We might be more comfortable, as scientists and clinicians, with facts. We might be unwilling to think or talk about moral questions, preferring just to repeat whatever we have done before or do whatever others tell us, or to avoid making decisions in the hope that the situation will somehow get better anyway. Indeed, we might not need ethics if we had no morals, did not have to act in the real world, or had a complete set of strict, irrefutable protocols initiated by specific evidence.

However, we have professional responsibilities in a complex and uncertain world (not least since COVID-19), which means we need to make professional judgements. As veterinary professionals, we do not blindly obey textbooks, rely only on our intuitions, wash our hands of difficulties, or dismiss veterinary topics as mere matters-of-opinion because there are disagreements. Instead, we think carefully about each case, make responsible judgements, and continuously develop our skills. So too with ethics. Ethics can help us to be more confident in dealing with ethical conflicts, avoid later remorse or anxiety, and reduce our overall stress levels in practice. It can also help us to discuss our views with clients, colleagues and students, in order to improve mutual understanding, to constructively challenge and defend one another, and to reach agreements.

Ideas

 Too clever by half
Sometimes being sceptical can make us feel clever or superior, but prevents us learning helpful new ideas or approaches. Instead, we should be open to approaches and fields outside our comfort zone.

 Ethics isn't nice
Sometimes we feel morally uncertain, perplexed, challenged, stressed, powerless, guilty or indignant. For caring professionals, unpleasant feelings are an unfortunate and undeserved aspect of the job, but can help us develop.

 Self-confidence
Sometimes we feel unwilling to reflect on what we do, or have done before, for fear of feeling stupid or guilty. Occasionally, we feel overly defensive (like a sort of moral hypersensitivity), self-destructive (like a moral autoimmune disease) or overwhelmed (like a sort of moral Disseminated Intravascular Coagulation (DIC)). We need enough confidence to implement our ethical views assertively, but be open to questioning our preconceptions without defensiveness or bluster.

 A bit too quick
We might have immediate thoughts about a case or jump to a solution that sometimes misses other issues (like treatment side-effects). Instead, depending on urgency, we should consider all relevant issues and sensible solutions.

 But what do we actually do …?
Some cases make us feel sadness, anger, indignation or despair. While these might be justifiable feelings, we should not also feel guilty that we cannot perfectly solve problems due to other people, natural processes or chance. Ethics can help us focus on the question of what we can and should do. (It can also help us understand others' behaviour, which might, partly, assuage our anger.)

Reflections

Reflecting on the cases we have experienced:

- What cases commonly occur in our own work?
- What cases do, or might, we find morally difficult or challenging? When have we felt confused, worried, distressed, overwhelmed or guilty? When did we lack confidence in our decisions?
- When were we completely clear and confident about what we should do? Were we ever too confident? Should we challenge ourselves more?
- When did we feel we knew what we should do, but still faced communications difficulties or emotional pressures because of others' moral views? When do we struggle to defend our views to other people?

For a particular case:

- Did we take responsibility for making decisions, or pass responsibility to someone else, or avoid making decisions at all?
- Did we focus on our decision, or spend (too) much time or effort thinking about how the world could have been different (e.g. lamenting owners' failings, wishing we had additional knowledge or equipment, or wishing we could choose impossible options)?
- Did we explicitly consider the ethical aspects of our decision-making, or just focus on the facts (and, if so, what do those facts tell us about our background ethical assumptions)?
- Were we too quick to make a decision or too slow? Did we make a decision conscientiously or leap to the first solution that presented to us?
- Was the decision easy or hard? (If so, what made it so?)
- Did we consider a wide range of ethical views that could be relevant, or simply approach the case the same way we always do?
- Did our ethical thinking help us? Did we come to an actual decision that we actually implemented?
- Did others share our views and agree with our decision? (If so, why? If not, why not?)
- Are we open to changing our views?

Part A: Understanding Ethics in Veterinary Practice

In this section, we consider key ethical questions. In each chapter, we will recall some of our own (i.e. each reader's) cases that we have seen in our own work or work experience, asking important questions in relation to them, and suggesting possible answers to those questions (i.e. 'We might think …') and why we might agree or disagree with them.

In this section, we should each keep formulating, analysing, challenging and reformulating our views, so that we end with *provisional* views that are defendable and applicable (depending on what other factors might also affect specific applications). These views are, at each stage, provisional but incomplete views, but we may think them right to some extent, all else being equal.

Considering Others **2**

A key part of ethics is about how we treat others (indeed, ethics might seem ignorable if nobody else existed). In this chapter, we consider how our behaviour can, and should, affect other animals and people. In section 2.1, we start by considering their experiences and motivations. In section 2.2, we then consider how we should deal with uncertainty in our predictions of outcomes. In section 2.3, we consider how we should deal with situations where we might affect multiple different animals and people.

2.1   Achieving outcomes for someone else

Our behaviour affects lots of animals and people: our patients, clients, colleagues, other animals and other people. We can cause them suffering or enjoyment, satisfy or frustrate their motivations, and affect their abilities to make and implement decisions.

Let us recall cases where our decision had an impact on an animal (e.g. on its medical, surgical, behavioural or environmental management) or a person (e.g. on their finances). How should we think about those outcomes?

- What outcomes should we avoid or aim for? How should we combine multiple possible outcomes for someone?
- Should we help (and avoid hindering) other people in making and implementing their decisions (and, if so, how and when)? (How) does this apply to animals?
- How should we combine our concern for outcomes with our concern for others' motivations?
- Is helping people to be better part of helping them?
- Should we try to protect life regardless of its quality? Should we consider whether animals want to live?

Table 2.1. Outcomes to consider.

Outcome	Examples
Specific outcomes to avoid	Extreme pain
	Maternal separation
	Starvation
General outcomes to avoid	Pain
	Anxiety
	Frustration
Overall outcomes to minimize	Suffering's intensity (mild/severe)
	Suffering's duration (acute/chronic)
	Suffering's frequency (low/high incidence)
General outcomes to achieve	Pleasure
	Satisfaction
Overall outcomes to maximize	Pleasure's intensity (extreme/mild)
	Pleasure's duration (sustained/transient)
	Pleasure's frequency (high/low incidence)
Overall outcomes to balance	Pleasure 'minus' suffering

We might be concerned with avoiding particular outcomes for others (Table 2.1). We might be concerned about particular extreme harms (although their definition might seem somewhat arbitrary), or that we should avoid or at least minimize harms in general. More positively, we might also aim to achieve, or maximize, certain outcomes (e.g. pleasure), although we cannot always help everyone to have completely enjoyable lives.

When a patient's or a client's outcomes will involve both suffering and enjoyment, we might think we should always prioritize the prevention of suffering, but this could suggest that we should kill every animal. Alternatively, we might weigh up all future suffering and enjoyment together, and choose whatever option we evaluate as having the best 'overall outcome' (e.g. 'subtracting' the total suffering from the total pleasure). In practice, we might sometimes find such composite calculations to be overly complicated, or downright impossible, to do confidently, although we might feel able to make some useful comparisons in some cases (e.g. that effective analgesia reduces postsurgical pain or tail-docking is more painful than flystrike or being tail-bitten).

*

We might think we should let, or help, people make and implement their own decisions (Table 2.2). However, people cannot always make perfect decisions (or at least I don't). We might think we should overrule people when 'we know best', but we might not feel confident in claiming that knowledge for other adult humans (i.e. they know what's good for them). Alternatively, we might think we should still respect someone else's choice even when we think it will lead to bad outcomes for them (e.g. letting clients

Table 2.2. Reasons to respect others' decisions.

Reason	Factual assumptions	Possible exceptions
Self-determination is vital for human flourishing or wellbeing	Self-determination is part of how humans should be	Decisions that damage third parties' flourishing or wellbeing
Allowing everyone to do what they want would generally lead to the best outcomes overall	Everyone (usually/ always) knows what is beneficial for themselves and makes decisions on that basis	Where people have limited knowledge or poor decision-making
We would want others to respect decisions and should be consistent	What we want to do is morally acceptable	Where we think others should stop us doing what's morally wrong, and vice-versa
Rational decisions deserve moral respect in themselves	Humans' decisions are rational	Human decisions that seem irrational (or non-rational, e.g. merely emotional)
People should be able to decide what happens to their own minds, bodies or possessions	People have the ability to make and implement decisions	Decisions that also affect others' minds, bodies or possessions

choose how much money to spend) – unless their choice would harm others (e.g. not letting clients choose to abuse animals).

In comparison, we might think we should not respect the decisions of non-human animals. However, it seems overly simplistic to believe all humans are always right and animals are always unreliable (e.g. both may be able to predict things that will hurt them). Alternatively, we might think we should respect the decisions of both humans and other animals, albeit sometimes in different ways (Table 2.3). We might think we should respect animals' decisions when we believe they can make sensible choices (e.g. in line with their evolved or learnt abilities), but not where we think they lack the relevant ability (e.g. medical decisions) or would choose to harm others (e.g. through aggression). We might think some animals should be left completely unrestricted (e.g. healthy wild animals in their ecological niches).

We could combine our concerns for outcomes and for animals' decisions by aiming for what animals would have wanted if they could understand and decide better. We might attempt a sort of 'preference dialysis', to refine and recreate what choices they prudentially would make, if they could. This approach would take into account our knowledge of the individual and of other animals (Fig. 2.1). For example, we might respect a diabetic cat's evident desire to avoid hospitalization *and* its implied preference to avoid suffering from ketoacidosis (which we assess requires treatment), as a 'refined' preference for us to monitor stability via urinalysis.

Table 2.3. Ways to respect the decisions of humans and other animals.

	Restrictions	Empowerment
Making decisions	Lying and undue influence Behavioural manipulation based on genetic modification, developmental restrictions or punishment	Providing information, education and professional advice Facilitating experiential and social learning Desensitization and counter-conditioning. Providing relevant (e.g. natural) stimuli/cues
Implementing decisions	Space or behavioural restrictions; habitat damage; bodily mutilations Coercive restrictions (e.g. rollkur)	Providing resources Providing environments that resemble their ecological niches

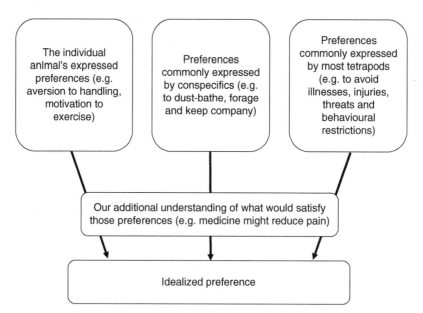

Fig. 2.1. Preference dialysis.

For humans, we might go further and think that acting ethically is not something separate from someone's self-interest, but a vital part of our personal flourishing. In other words, what is best for a person is for them to be moral. Negatively, this suggests we should not help people where doing so would help them to be immoral. Positively, this suggests part of our ethical responsibility is to help others to be ethical. We might therefore try to work out not only what people want, or would want, but what they should want – as 'morally refined' preferences.

*

We might also think animals' lives are worth creating or protecting. We might feel life has value in itself (and we might apply this to humans, to some animals, or to all living beings). However, this view could conflict with our concern to avoid suffering (e.g. whether to keep suffering animals alive, or even create them). We might therefore always prioritize life over suffering (which would suggest never euthanizing suffering animals) or suffering over death (which suggests we should kill all animals at birth). Alternatively, we might avoid such conflicts by saying that life has value depending on the quality of that life. This suggests we should create or extend an animal's life if, and only if, it would be enjoyable or the animal wants to live, and end lives expected to involve more suffering than enjoyment.

We might infer whether animals want to live from their behaviour (e.g. eating and hiding). However, we might think most non-human animals lack the cognitive concepts (e.g. of mortality and 'oneself extended over time') that are needed for explicit existential preferences, and that their behaviour actually relates to different motivations (e.g. hunger and fear). Alternatively, we might think animals' multiple preferences for what life will include can be refined into a preference for life or death. This would suggest animals would choose to extend an enjoyable life but to avoid any 'life-worse-than-death'. (We might also think that most humans can have preferences for life or death, but this is not a particularly salient question in everyday veterinary work, outside of preventing fatal zoonoses and resisting the urge to murder some clients).

Ideas

 It's good to be alive (sometimes)
We might say some lives are worth living for the animal, whereas others include such suffering that they are worth avoiding, either by improving the life or by euthanasia.

 Ethical associations
Some states or things might be said to have value in themselves. Others are valuable because they are associated with things that have value, i.e. they cause or are caused by valuable states. We should avoid treating something with associative value as if it is valuable in itself (e.g. medicine and research).

 Misdiagnosed guilt
We might feel experiences related to others' outcomes, such as sadness, anger and sympathy. These might help our decision-making, but we should not confuse them with guilt for what we have wrongly caused. We may feel bad for a patient but this does not mean we are responsible.

Reflections

- What would cause suffering, frustration, enjoyment or satisfaction for this patient?
- What do we believe (as best we can) this animal would want to achieve or avoid (and how strongly), given its species and signalment, and its individual behaviour?
- Might we be imposing our own preferences on other individuals or species (e.g. to have progeny, produce profit, generate data or win competitions)?
- What do we think this patient would want as a 'refined' choice, if it kept its basic psychology but could also understand medical options, and make and express medical decisions?
- What would our client want?
- Can we help the client make a better assessment of what they want and the impact of each option on them (e.g. if they want a healthy animal, we can help them identify what management would achieve that)?
- Should we provide euthanasia (and not help breed) for animals who are expected to suffer lives-worse-than-death, or is all life valuable?

2.2   Dealing with uncertainty

Most ethical decisions involve predictions about the future. However, in practice, we are faced with biological variability, other people's unreliability (whatever they promise), and just simple chance variation. We rarely have all the data we would like; even published data represent statistical uncertainties extrapolated from other cases, and some activities have almost completely unknown risks (e.g. novel procedures).

Indeed, we never even know for sure that we will cause the outcomes we intend. Sometimes, we merely change the statistical chances of an outcome (which may or may not have happened anyway). Sometimes we try to achieve one outcome but cause unlikely, unintended or unexpected side-effects (e.g. drug reactions and perioperative mortality). Even when we are looking backwards, we do not know what would have happened if we had done nothing or done something different.

Let us each recall cases when we were unsure about how different options would turn out. How do we decide what is best when we are faced with uncertainty?

- Looking forwards, should we try to increase our certainty or leave matters to chance? Should we ignore probabilities and just focus on avoiding or achieving particular outcomes? Should we take probabilities into account? If so, how?
- What evidence should we use to predict probabilities? How factually confident should we need to be? Should we use non-scientific evidence too? Should we ever be deliberately biased?
- Looking backwards, should we hold ourselves and others accountable based on the outcomes of their behaviour?

We might think we should try to minimize uncertainty by selecting more controllable interventions (e.g. hospitalization may eliminate variability in owner compliance), although we cannot remove all uncertainty (unless we kill every patient). We might think we should leave outcomes to chance (e.g. owners might prefer to risk their animal dying than to actively choose euthanasia), or even ignore likelihoods completely (e.g. owners may want treatments with miniscule chances of success and possible side-effects). However, we might think we are still responsible for outcomes that we knowingly allow, when we could and should have prevented them. (In practice, we cannot just blame God.)

So we might think we should take likelihoods into account. We could think we should only choose options with at least a 'plausible' likelihood of success. However, this faces the challenge of working out how probable an option should be for us to permit it (e.g. 5% versus 50%). We might believe that we should base our decisions on the statistically optimal 'best bet' option across all possible outcomes, factoring probability alongside intensity and duration. However, this makes our decisions even more complex, depending on what relevant statistical data (if any) are available.

*

A related question concerns what evidence we use. As scientists, we often use continuous and probabilistic data, and usually avoid definitive conclusions (e.g. rejecting null hypotheses only if we have statistically significant data). This feels fine and healthy as researchers. So we might be similarly tempted to delay making ethical decisions until we have certain evidence. However, we do not usually have that luxury in practice. We have to make decisions – otherwise we effectively decide to do nothing. Indeed, our ethical decisions have to be definitive (i.e. what to do now) and categorical (i.e. to do this or do that).

*

A related concern is what amount and type of evidence we feel we need. For example, we might require scientific evidence of long-term physical harm before wanting to phase out a production system. However, this can lead to delays (while animals suffer), us being seen as weak or complicit, or even a disincentive for funders who favour the status quo or laissez-faire approaches. It can also lead to our concerns being outweighed by those of other people who do not have the same statistical strictness (e.g. owners or lobby groups), or where we accept non-scientific evidence (e.g. economic data).

So we might think we should tailor what evidence we require to each decision. We might try to use the best scientific evidence available, but also be open to using non-scientific evidence where morally justified. We might think we should make initial ethical decisions (e.g. based on our expert confidence or a 'balance of probability') and then remain open to revising them as we gain new data (e.g. as cases progress, or using Bayesian statistical methods). In particular, we might 'err on the side of caution' and limit the use of new interventions or systems until they have been adequately tested.

*

Looking back, we might think we should hold people morally accountable depending on what outcomes they have caused or allowed – just as owners might say vets were wrong for advising euthanasia if their animal then recovers (and somehow it's always the most difficult owners). However, we might feel it is unfair to hold us accountable for being unlucky. Alternatively, we might think that our moral accountability should be based on whether we increased the likelihood of bad outcomes, which means a decision might have been right even if it turns out badly (or, conversely, wrong even though it turns out fine). However, this makes our moral concerns seem less clear and hard to assess – we cannot be judged on our results. It also makes our ethical motivations problematically broad: it is easier to follow 'do not kill' than 'do not create a significant left-shift in the statistical distribution of mortality'.

Ideas

 Hoping for miracles
Sometimes we (and owners) fail to recognize that some treatments have a negligible chance of the desired outcomes, relative to no treatment. We should instead consider all risks proportionately.

 The feel of risk
Probabilities can feel very different to different people (e.g. clients may be very concerned about small chances of mortality or a miracle cure and want to 'try everything'). We should help people consider risks proportionately, 'steady our nerve' to provide the 'best bet' for our patient when we are unsure how it will turn out, and 'hold our nerve' to avoid pressure to choose heroic options with low odds of success.

 Hindsight is always 20/20
Our same decision can feel, and look, very different depending on whether we are lucky or unlucky. When choices turn out badly, we should reflect on whether we should have made a different decision at the time (with only the information we had then) or were just unlucky. If so, we should avoid criticizing ourselves, and be more resilient to others' misplaced criticism. (We should similarly avoid assuming we were right because outcomes turn out fine.)

Reflections

- What makes a decision right – the process or the outcomes (or both or something else)?
- Where can we get the most reliable information on patients' preferences (e.g. scientific studies, our experience of similar animals, owner insights)?
- For each possible outcome, what is its severity and likelihood?
- How confident are we that a given outcome would occur after each option? Are reported probabilities (possible as incidence measures) available? If so, is it valid to extrapolate to our patient? If not, what is our subjective confidence that an outcome would occur?
- Is the potential suffering so severe and its later control so unreliable that we should err on the side of prevention (e.g. euthanasia)?
- Should we categorically set a plan to a predetermined final outcome on the information available now (or is that too inflexible) or should we initially aim just for a more 'proximal' outcome first, and adjust as situations change and new information becomes available (or is that too 'dithery')?
- Are we being overly swayed by remote chances or a desire for certainty and control?
- What is the 'best bet' for each animal who might be affected?

2.3   Thinking of everyone

We have considered outcomes as they apply (statistically) to individuals. However, our decisions often affect multiple animals and people. What is better for one animal might be worse for others in their social group, population, species or ecosystem (e.g. spreading disease or deleterious genes); be suboptimal for their owners (e.g. reducing profit or competition success); create risks for members of the public (e.g. promoting antimicrobial resistance); encourage the use of other animals (e.g. in food or medicine production); or use up limited resources.

Let us recall cases where we predicted one option would help one animal but be worse for someone else. How do we resolve clashes between the interests of multiple animals and people?

- Should we always prioritize some humans or animals, or should we treat everyone consistently (if so, how)?
- Can we consider all outcomes in one overall assessment?

We might think we should always prioritize one population, such as humans (e.g. destroying animals with zoonotic conditions) or patients (e.g. using medicines tested on other animals). However, this approach could suggest

severely harming animals for very minor benefits to people (e.g. supporting cosmetic testing or bear-baiting) or vice versa. We might also struggle
to justify any such prioritization morally without it simply seeming like a
fundamental personal bias (e.g. towards our own species). We might instead
think we should consider species (and other factors such as gender, race and
age) only insofar as they correspond to morally relevant factors (e.g. diagnostic signalment).

Alternatively, we might think we should treat everyone consistently.
However, while this sounds fair enough, it is hard to work out what is consistent, because what is consistent in one way may be inconsistent in another
(e.g. giving everyone the same resources may mean each has different outcomes), especially as different species and individuals have varying needs and
preferences (e.g. most humans do not enjoy rooting, rolling or licking themselves). We might also think that ensuring complete consistency is impossible, at least without excessive interference (e.g. controlling all wildlife) or
reducing everyone to some lowest common state (e.g. death).

Less ambitiously, we might set some consistent limits on the worst outcomes we allow any individual to suffer (e.g. never allowing severe suffering,
starvation, or a life-worse-than-death). This would require some defence of
what that minimum is (and why it is not higher or lower), for example based
on the best possible minimum that could be achieved. However, this does
not help decide what to do in cases where we cannot prevent some animals
being below that standard (e.g. wildlife or laboratory animals) or what to
do above this minimum. We might then combine this approach with a different approach (e.g. the one below), but that raises the question of why not
simply use that approach alone.

*

A very different approach could be to treat all involved as if they were one
patient, and doing what is likely to cause the best outcome overall. Any
harms to some individuals may be 'outweighed' by greater benefits to others,
regardless of who benefits or suffers (and regardless of their species). This
effectively tries to work out what all members of the group would choose if
they could understand medical decisions, but were ignorant of who would
suffer which outcome. However, when applied to all animals and people
worldwide, now and in the future, we might find that this approach suggests some challenging conclusions (e.g. that we should be mainly vegan,
use humans for medical experiments, steal money from rich clients to help
poorer ones).

Even if we agree with this approach in theory, we then face a profound
practical challenge of 'adding up' the values of everyone's outcomes. We
might think it is impossible to accurately compare and combine all the outcomes for everyone affected (e.g. the impacts of an intervention on a whole
herd, the farmer, future animals on the farm, consumers, other farmers, the
market, the wider economy, the climate). Nevertheless, we might use this

method to simplify some decisions by discounting benefits (or harms) that come with equal or greater disadvantages (or benefits) for others (e.g. that profit, aesthetic preferences and sporting enjoyment are outweighed by major injuries, disease or confinement). We might also simplify some calculations by ignoring effects that cancel each other out (e.g. relative competitive advantages for a farmer, breeder or trainer) or seem negligible (e.g. very minor indirect harms). But this approach still seems daunting in practice.

Ideas

 Greatest good for the greatest number
We might think we should always do whatever has the best outcomes, based on an evaluation of the totality of outcomes for everyone.

 What might have been
We might recognize that each option not only has disadvantages in itself but also means foregoing the relative benefits of the other options.

 Cancelling out
Sometimes benefits to one person or animal would be equivalent to others' losses, so there is no benefit overall (e.g. in gambling, competitions, some financial derivative returns).

 You can't please all of the people all of the time
It can feel sad when we cannot help everyone, and unfair on us to have to choose whom to help. We should avoid feeling guilty for not helping everyone (even if they complain) when we do so to help others.

Reflections

- Are clashes unavoidable or can we make one option best for everyone (e.g. helping clients to want what is best for their animal)?
- Does any option create an outcome for one individual that we cannot justify by benefits to others? Does any option go beyond the limit of how much suffering any individual should endure? Does any option mean any individual has a life-worse-than-death?
- Are the benefits to some animals or people outweighed by greater disadvantages for others (or vice versa)? Do any effects cancel each other out? Can any be ignored as negligible?

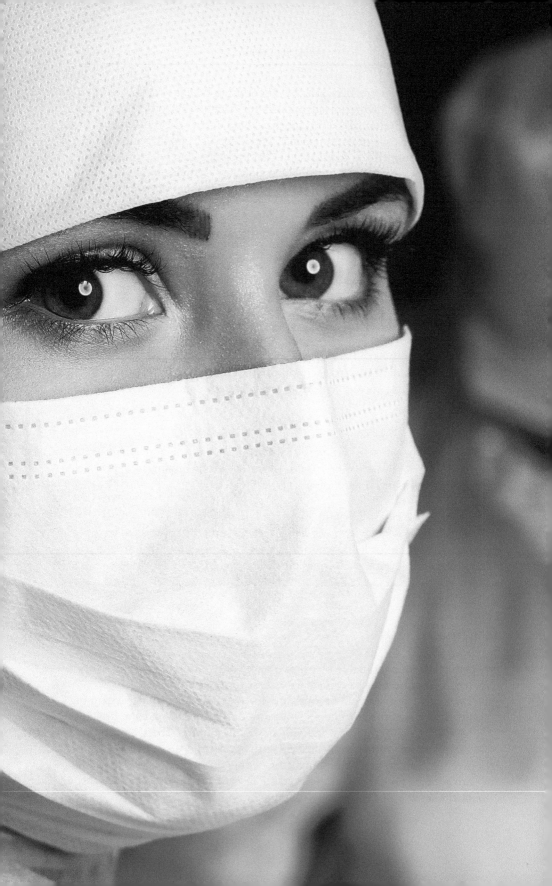

Considering Ourselves

3

So far, we have considered others. We have ignored ourselves. This chapter considers ways in which we each feature in our ethical decisions: how our decisions affect ourselves, and thinking of ourselves as moral practitioners.

3.1 Our personal outcomes

> Veterinary work can be rewarding or stressful; gain or lose us money, time and popularity; satisfy our intellectual curiosity; get ourselves sued; and make us feel like a hero or a failure.
>
> Let us recall cases where some of the options would have benefited or harmed ourselves. How should we take our own outcomes into account?
>
> - Can we think of our outcomes like anyone else's? How selfless should we be? Should we avoid any option from which we would benefit?
> - Should we consider the feelings that our behaviour might cause ourselves? Should we try to ignore our emotions? Do our emotions undermine the credibility of ethics?

We might believe we should be as concerned with outcomes for ourselves as for anyone else (e.g. our income is just as important as that of non-subsistence farmers), albeit recognizing we are each only a single person and being careful not to pay disproportionate concern. Conversely, we might believe ethics is all about others, so we should ignore personal outcomes. However, this might suggest we should engage in extreme self-sacrifice, even when doing so harms us more than it helps others. Combining these views, we might think we should ignore personal benefits and minor harms, but limit what personal sacrifices we should be obliged to make (e.g. insolvency or mental exhaustion), even if making them is laudable.

We might also recognize that our benefitting does not necessarily mean an option is wrong. We can help patients and clients while charging a fair fee (indeed commercial veterinary practice business models generally assume that both transacting parties benefit), and our career development might help future patients. Similarly, some harms to ourselves are associated with risks to others (e.g. we cannot help future patients if we are dead, injured, mentally unhealthy; we lose our legal licence to practise; or our practice is unsustainable). Even if we ignore ourselves directly, we should still consider these sequelae.

*

Another way in which we are affected by our decisions is through our moral feelings. We might feel distress, anxiety or self-pity when we feel unable to know what is right to do, or frustrated when we cannot do what we want to do morally. We might feel a 'warm glow' of satisfaction for having done what we think right (although we may often miss this latter feeling because we feel too bad about the situation, or feel taken advantage of by a client). We might feel weak or guilty for having done something morally wrong, or shame at being morally criticized by others (and sometimes we feel guilty even if we did the right thing in a bad situation). We might also feel approval or indignation at others' behaviour or statements.

Indeed, there is an argument that ethics is, ultimately, all about emotion. Unless we are simply and impassively comparing options to predetermined rules, we inevitably need to make evaluative (i.e. value-based) moral judgements. These are, in one sense, emotional judgements. We might prefer or dislike one option. We might find one idea disgusting, repugnant or undignified. We might desire – for moral reasons – to achieve one option or avoid another. Psychologically, ethics probably has to involve some emotional aspects to be motivational and meaningful for us. A purely fact-based algorithm could not make moral decisions for us (unless we build our concerns into the underlying logic of the algorithm).

We might worry that this emotional aspect undermines ethics, insofar as it is not purely a logical or objective exercise. Alternatively, we might say this is simply a recognition of our psychology: we do not reject the ideas of 'love', 'beauty' or 'pain' because they have emotional components, nor deny the existence of the universe simply because we only know about it through our sensations. Nor should we dismiss ethics because our appreciation of it is linked to our emotions and evaluations. We could think of morality as a sort of refined motivation, which considers everyone and excludes what we should want not to want. We might say that, as vets, we should morally want what our patients want (and we should want our clients to want what our patients want).

Ideas

 Selflessness

Some views and behaviours are based on a selfless concern for others, without any expectation of reward, recognition or reciprocity (even if

continued

those benefits do then come as side-effects). Even where we might have evolved such selflessness, this does not make it any less selfless.

 Predisposed to help ourselves
Sometimes our moral decisions might be subconsciously biased towards what benefits ourselves, especially as we are more aware of our needs than those of others. We should be aware of our interests explicitly, to help us be impartial or selfless.

 Feel-good factor
There is a positive feeling associated with helping others (even if we sometimes forget to feel it in the midst of daily struggles and pressures).

Reflections

If we decide that we should usually ignore outcomes for ourselves in our clinical decision-making, we might still ask:

- Would any harm to us also genuinely stop us helping others in future?
- Might we be subconsciously biased towards outcomes that benefit ourselves?
- What about a situation that is making us feel sad, distressed or anxious? Can we address those emotions before we make a decision, or do we need to make a decision to address them?
- Are my emotions helping or hindering my decision (or both)?
- How would (and how should) our different options make us feel (e.g. guilty or satisfied)?
- How do (and how should) I feel about what I have done in the past?

3.2   Our personal moral phenotype

We each have personal moral character traits, perform particular behaviours and have specific effects. Often we are faced with cases where what seems best is something we would rather avoid (e.g. euthanizing healthy animals, passing on client data, or 'mutilating' animals to prevent harms that could be avoided by different management) but the situation makes it seem the best option (e.g. if an owner threatens to drown the animal or plans to continue harmful management). We also can cause unintended outcomes; for example, veterinary management can cause experiences such as anxiety, frustration, pain and loss for our patients and clients.

continued

Let us recall cases where what seemed best for our patient involved us doing something we would think is generally wrong. How should we think of ourselves as moral practitioners?

- What traits should we show? How, and how much, should we show such traits (and when and to whom)? Does it depend (if so, on what)? How should we work out the right traits to display?
- Should we place any particular importance on our behaviour or traits, regardless of their side-effects or outcomes we could have prevented? Should we stick to our guns or pragmatically do whatever achieves the best outcomes?

Table 3.1. Moral traits.

List	Traits
Brahmavihārās	Goodwill, compassion, sympathetic joy and equanimity
Cardinal virtues	Courage, wisdom, temperance and justice
Theological virtues	Faith, hope and charity
Seven deadly sins	Pride, envy, wrath, sloth, greed, gluttony, lust

We might think it important to possess and develop traits such as kindness and courage and avoid others such as cruelty and idleness (Table 3.1). However, we might find such traits seem too imprecise to help us make concrete decisions in particular cases. We might struggle to work out the right combination of traits and to what degree we should display each one, especially when different traits seem to suggest different responses (e.g. how to combine courage and prudence), or when the same trait has internal conflicts (e.g. compassion when the needs of patients and owners conflict) or seems to vary depending on each person's role (e.g. an orthopaedic surgeon might need different traits to a veterinary nurse) and circumstances (e.g. in a clinical emergency versus a performance meeting with an employee).

To resolve these difficulties, we might identify optimal traits as the midpoints between pairs of opposing undesirable traits (Table 3.2). We might try to emulate moral exemplars (e.g. Jesus or Gandhi). We might 'convert' traits to considerations of behaviour (e.g. 'honesty' to 'not lying') or outcomes (e.g. 'compassion' to 'preventing suffering'), but then we might as well just think directly about behaviour or outcomes, and ignore traits. We might choose general traits that achieve some further end (e.g. others' pleasure or our own flourishing) or that help develop other traits (Table 3.3).

We might also be particularly concerned with our own moral behaviour, regardless of its outcomes. There may be some behaviours that we just feel are wrong, even when doing them would seem harmless (e.g. 'white lies' to clients) or even when they would prevent the same or worse outcomes due to another aetiology (e.g. lying to clients to prevent further animal abuse). We might be particularly concerned with outcomes that we cause rather than allow, or that we intend rather than cause as side-effects (e.g. providing life-saving treatment regardless of the subsequent suffering). However, this obviously risks us causing harm and missing chances to help others. We might also think such distinctions dissolve when we have undertaken to care for animals (e.g. in clinical practice).

Alternatively, we might think we should not place any particular emphasis on our own role in the process by which outcomes occur. We should make pragmatic decisions based on whatever reality is, including others' behaviour, just as we consider other factors (e.g. co-morbidities). However, this approach suggests that any behaviour could be legitimized in certain extreme cases (e.g. abusing clients to prevent even worse animal abuse).

Table 3.2. Traits and opposites.

'Hypo'	Optimal trait level	'Hyper'
Stinginess	Generosity	Prodigality
Envy; Vindictiveness	Mercy	Weakness
Dishonesty	Honesty	Indiscretion
Inflexibility	Empathy	Complaisance
Timidity	Courage	Recklessness

Table 3.3. Some traits that might underpin ethics in general.

Practical wisdom to help us choose traits
Commitment to keep developing them
Openness to debate and being personally reflective
Skill and conscientiousness in our decision-making
Courage and resoluteness in implementing our decisions in the face of pressure
Selflessness and impartiality

Ideas

 Let or cause
We might feel more responsible for what we do (e.g. murder) than for what we omit to do (e.g. not saving a life), even if the outcomes are the same. Alternatively, we may feel responsible when we fail to act in ways we think we should.

 Iatrogenic side-effects
We might think we are not responsible for unintended consequences of our behaviour, as long as we did not intend them (even if we foresaw them) and they are not integral to our behaviour or disproportionately large.

continued

Rose (or other) coloured spectacles
Sometimes we approach problems as if the world were perfect (e.g. compassionate clients and perfect communication) or awful (e.g. clients are completely immoral). We should instead tailor our decisions – and make contingency plans – for what we predict can or will realistically occur.

 Self-worth
It is important to feel that one is 'a good person' and horrid to think of oneself as bad. However, we should not think of ourselves as morally perfect either, but always be motivated to get better.

♥ *Feeling dirty*
When we do something that we think is wrong in order to prevent worse outcomes, we may feel resentful or indignant against those whose behaviour has caused us to face that dilemma and then afterwards self-contemptuous for being 'weak', complaisant, 'dirty', compromised or corrupted, and bad – paradoxically – for doing what we thought was right. Conversely, sticking to our guns might be difficult at the time, and afterwards might feel self-indulgent or egocentric.

Reflections

Thinking about our own phenotype, we might ask:

- What set of characteristics should I exhibit?
- How much should I show those characteristics?
- Am I only or overly focusing on one trait, and ignoring or downplaying others?
- Would displaying a particular trait make me feel comfortable with, or proud of, myself?

For a given behaviour:

- Is this a behaviour that we should always/never do even when/if it would help a patient in exceptional circumstances? Can we think of any exceptions?
- If we think we should do something purely because of another person's immoral behaviour (past, present or future), can we change their behaviour?

Considering Our Relationships   **4**

So far we have considered ourselves and others separately. However, we are not isolated, unconnected individuals. We have various relationships with particular animals and people, such as our patients, clients, animals we own, family members, colleagues, employees, employers, regulators and wider society. These relationships can feel morally important, especially those with a degree of intimacy, familiarity, vulnerability, dependency, reciprocity or expectation

In this chapter, we consider how our ethics might be more specific to our particular relationships (in the same way that we might think our relationships with our spouses and family should morally affect our interactions with them).

4.1   Patients, clients and us

Some animals are our patients (whether they want to be or not). Some people are our clients (whether we want them to be or not). Patients under our care are dependent on us for their needs to be met. Clients rely on us for veterinary services that they cannot, or are not allowed to, provide themselves. Society expects us to protect and help animals on everyone's behalf. We may have also made specific undertakings (e.g. in making promises, signing contracts, accepting clients' money, subscribing to practice policies, or swearing professional veterinary oaths to care for animals). We interact with them in the context of these specific interpersonal relationships.

People and animals also have significant relationships with one another. Clients often legally own our patients and effectively control how (and whether) they live. Many manage their animals well and want to do even better. Others are less perfect – to varying degrees, through lack of capabilities (e.g. understanding) and opportunities (e.g. decision-making skills), or misdirected emotional or ethical views that can harm animals (e.g. if an owner's attachment stops them allowing euthanasia or leads them into hoarding). Indeed, many – if not most – of the problems we face are ultimately due to what owners have

continued

done or want to do. To make it even trickier, sometimes helping the animals can allow owners to continue poor animal care (e.g. providing corrective surgery for inherited disorders in breeding animals, or providing routine antimicrobials or preventive 'mutilations' for systemic problems).

Plus, animals may have relationships with each other, in bonded groups or families. Many animals show affection, empathy, loyalty and kindness, and suffer during social isolation or maternal separation (and some have apparent concepts of justice). Sometimes, these relationships are enjoyable or useful; but animal interactions can also be antagonistic or competitive, and even positive relationships can create conflicts between what is beneficial for different individuals (e.g. when a lactating queen or bonded rabbit will suffer without euthanasia).

Let us recall any cases that involved patients and clients, perhaps when the problems were caused by an owner's previous or expected management of their animals.

- Do we have particular responsibilities due to our relationships with our patients and clients, and how should they be affected by those clients' relationships with their animals?
- Have we made undertakings that we should fulfil – even if doing so seems otherwise morally wrong?
- What are our responsibilities to our patients or clients? May we morally harm them? Should we also particularly help them? Do we have these responsibilities to all animals/people or only to patients/clients?
- Do we have the same responsibilities to our patients and our clients?
- What should human–animal relationships be like? What control should owners have? What responsibilities should they have, and how should these limit their control?
- Do owners' responsibilities affect what responsibilities we have? If so, how?
- How should we consider relationships between different animals? Should these affect how we decide what is best for them?

We might think we should fulfil our voluntary undertakings, since otherwise we would have been absurd in making them, or because these undertakings increase our opportunities to help animals (e.g. by encouraging owners to present their animals and give honest histories) and breaking them would reduce these in future. However, if the best option changes (e.g. if more information comes to light or cases progress), we might think it equally absurd to stick dogmatically to undertakings that subsequently appear foolish. Alternatively, we might think that we should fulfil our undertakings only when we still think they are right. However, this weakens the value of such undertakings (e.g. reducing our reliability for others), unless we make the caveat explicit (e.g. by promising to do something 'unless circumstances change').

*

We might think we should not harm our patients or clients (i.e. never cause outcomes that are worse than doing nothing). When we stop animals from meeting their own needs, we should ensure their outcomes are no worse than they would otherwise have been. When we keep patients alive, we should minimize subsequent suffering. When we legally limit veterinary practice to ourselves, we should provide better care than lay people would have done. When we claim that our views are authoritative, we should ensure we have the expertise (scientific and moral) and are sufficiently progressive. However, we might feel that avoiding harm still biases us towards inaction (we could avoid all harm by never doing anything).

So we might also think we should help our patients and clients positively, in ways that do not necessarily apply to other animals and people (e.g. we should give analgesia to hospitalized patients but not necessarily fund international veterinary charities or roam the forests looking for injured squirrels). We might even think we should help patients when this risks harming others (e.g. rehabilitating predators). Alternatively, we might think we should help our patients and clients only where this is possible without significantly harming others or breaking any other moral rule (e.g. we cannot eliminate all wild animal suffering without depopulating, which would seem disproportionately harmful).

*

We might think we have different ethical responsibilities to our patients than to our clients. We can claim expertise about animals but might reckon clients can be expected to know their interests best (however unintelligible we might think their preferences are) and we just provide technical services they want. We might think owners should be able to control what happens to their property (but not that animals should control what happens to themselves). We might believe owners' choices will generally lead to the best outcomes for those animals, although this seems naïvely optimistic. However, this logic could suggest we should let clients harm their animals or others – or even help them to do so.

Alternatively, we might think we should directly safeguard our patients' interests and not respect owners' control or freedom where that is harmful or otherwise immoral. We might think owners' control should be limited by their responsibilities to care for their animals, and more generally to protect their consumers, employees and others. This suggests we should not allow or help owners to severely harm their animals. This includes not letting owners prevent us from helping their animals. Together, these suggest we should sometimes help animals even when this frustrates an owner's wishes (e.g. giving unsolicited advice, providing emergency treatment without consent, or reporting suspected animal (and similarly child) abuse). Positively, we could see this as helping our patients to flourish and our clients to flourish morally, i.e. develop their ethical traits and behaviour (albeit only on matters where we feel able to judge these).

*

We might think owners' responsibilities should limit ours. We might think we are not obliged to fulfil owners' responsibilities when they decide not to (e.g. to feed our patients in their owners' homes, or to provide treatments that owners refuse to fund), and that the moral 'blame' for any resultant suffering is on the owner. However, this approach could effectively absolve us of responsibility to prevent any suffering in owned patients that their owners have caused or could prevent – which is possibly most of our cases, given how much owners affect their animals' lives. We might also think we should help some owned animals where their owners cannot fulfil their responsibilities for excusable reasons (e.g. incapacity or poverty).

Alternatively, we might think we and owners can both have the same responsibilities at the same time (like both parents have for their child – each cannot just abdicate responsibility to the other). In many cases, we can fulfil our responsibilities while they also fulfil theirs (e.g. they pay us for providing services). In other cases, we might need to fulfil our responsibility even though owners fail in theirs. This suggests we should pragmatically do what is needed, even when that means compensating for owners' failures. However, this can risk perpetuating poor practices, and potentially even lead to a spiral where our efforts to ameliorate problems allow further problems to develop that then need addressing (e.g. genetically or surgically modifying animals to 'fit' suboptimal husbandry systems can facilitate further management changes that need further modifications, or facilitating owners to acquire more animals that they cannot look after may mean they obtain more).

*

We might think we should consider animals' relationships with each other insofar as they affect their outcomes (e.g. ensuring maternal contact and care). When there are apparent conflicts between group members, rather than deciding between individuals (as in section 2.3), we might think of the group as having a communal interest that means what is beneficial for the group is beneficial for each (i.e. there is no conflict). In practice this might mean the same as considering overall outcomes, but feel less of a trade-off. However, while this approach might seem applicable to families, colonies, hives, flocks, herds or shoals (and, aquaculturally, we might even call it 'ethical schooling' insofar as all interests are aligned), it would seem invalid to extend it to animals connected only impersonally (e.g. of the same species or on the same farm).

Ideas

 Treating different ideas as the same
Sometimes we miss subtle distinctions between closely related concepts (e.g. 'humans/persons' or 'medical/veterinary patients'). This risks us unquestioningly transferring beliefs about one to the other (e.g. that veterinary clients should be treated like medical patients) without actually arguing why.

 Property
Property is a legal concept, but we might think it represents a legitimate ethical basis for owners having some control over what happens to their animals (and perhaps, by extension, human children or slaves).

continued

Covenants
In some cases, we might feel we or others have implicitly undertaken moral responsibilities (e.g. owners in obtaining and keeping animals, or ourselves in acting as veterinary professionals), because this effectively prevents others from acting and/or creates an understanding or covenant.

 Welfare arms races
Sometimes improvements that help animals to cope with their environments (e.g. tighter biosecurity) allow owners to make further changes that cause further problems (e.g. increased stress), needing new improvements to mitigate those welfare effects. In the end, the animals are no better off.

 Scrubbing up
Sticking inflexibly to our moral rules despite others' failings might feel like we are 'keeping our hands clean' despite the possible contaminants of others' ethics. Or it might feel like we are 'washing our hands' of a problem.

 Inherited guilt
Sometimes we take on others' guilt, as if we are personally responsible for problems caused by clients or society. We should instead feel sad or angry, but not guilty.

Reflections

- Do our relationships with our patients and clients create specific moral duties that do not apply to other people and animals?
- Is it ever right to harm patients to benefit clients, or other animals or people? If yes, does this mean we should just do whatever leads to the best outcomes overall, considering all animals and people as one patient impartially (i.e. with no special moral concern for our patients)?
- What should we do when we have undertaken to do something that later seems wrong?
- Do we have responsibilities to our patients that we do not have to our clients, and vice versa?
- How should owners manage their animals (as a minimum)?
- What (if anything) is different between the control a person should have over animal property, human property (which has sometimes included slaves, children and wives), insentient property (e.g. money) and their own body?
- (When and how) are we obliged to help animals avoid suffering that owners have caused or could prevent? Have they specifically delegated responsibilities to us (e.g. to feed hospitalized patients)?
- When owners cause or allow avoidable suffering, deprivation or frustration, (when and how) should we pragmatically 'bend' our moral rules to prevent or reduce it? Should it depend on the overall short- and long-term outcomes of each option? Should it depend on whether the harm is unavoidable, irreversible or excusable?
- (When and how) should we help (or not) animals if doing so unintentionally helps owners to continue their husbandry unchanged? Are there ways we can help animals now without facilitating poor management in the future?

- Should we consider a group of animals as having a common, collective welfare (e.g. members of a social group) or are the animals not personally connected (e.g. members of a breed)?

4.2   Our relationships with our colleagues

We also have particular relationships with our veterinary colleagues. We are members of a practice team, contracted employees, and members of a profession. What we do, and how it turns out, may also depend on what our colleagues will do (e.g. if our client might otherwise use another vet or we might get sacked for disobeying an instruction). In many countries, we are given our exclusive licence to practise veterinary work by society to benefit animals and humans and to manage risks in societal processes (e.g. pre-transportation checks for infectious diseases). We also receive requests or instructions from our bosses and regulators (e.g. practise standard procedures and professional codes), and some of us get to make those instructions (who might not always seem the best people to do so).

Let us recall cases in which our options were affected by what other veterinary surgeons or nurses did (or might do). How should we consider their views and behaviour?

- Do we have responsibilities to each other? Should what we do as individuals depend on what other veterinary professionals do?
- Do we have the same responsibilities as other members of society?
- Should we follow others' instructions? (If so, whose, when and how?) Or should we do whatever we would otherwise have thought right?

We might think our practices and professions are no more than populations of individuals, who should pursue their own interests, within agreed frameworks, and collaborate only where mutually beneficial. Alternatively, we might think we have specific responsibilities to one another, to which we should sometimes subordinate our individual interests, for example to avoid damaging our collective reputation (and ability to help animals and people), and to each 'do our bit' in fulfilling collective responsibilities (e.g. 'sharing the load' of providing emergency care for unowned animals).

We might feel our moral standards are more acceptable if they are widely shared, because we think morality depends on our culture, or because it would be unfair for us to be expected to go beyond common standards (e.g. if we thereby lose clients or profit). However, in emulating others we risk lowering our standards to those of the worst vets. Plus, what others do is logically irrelevant to whether we are moral (just as a patient is no less obese simply because other animals are fat). Alternatively, we might think we should emulate vets of whom we approve, maintain our standards where they are higher than those of others (i.e. not copy our worst competitors), and ensure we are no worse than the majority of vets (as that would disadvantage our clients and patients who come to us). We might also help improve one another's behaviour through mutual influence, advice and policy (and maybe even write a veterinary ethics book).

*

We might think we have the same generic responsibilities as any other members of society (e.g. to pay taxes), as it seems unreasonable to expect more of us (we are vets rather than saints). Alternatively, as veterinary professionals, we might think we should go further and follow a shared professional morality that is more than what is expected of lay people. We might think this 'extra' morality is simply general morality applied to veterinary specifics (e.g. not harming animals has specific implications for our work). Alternatively, we might think we have collective additional responsibilities, for example to care for animals especially because we have undertaken (and claimed) to care for them.

We might also think that, as a profession, we have collective responsibilities to human society. Indeed, we might think this is a condition of our exclusive licence to practise given by society, on the basis of societal expectations and undertakings (e.g. safeguarding public health, protecting animals, and legal compliance). However, this does not make us obliged to do anything unethical simply because an individual member of society (e.g. a client) wants it. Alternatively, we might think we have no obligations to society beyond any other citizen or business (e.g. not to evade taxes, harm animals, defraud clients or certify dishonestly). However, if we do not fulfil our obligations as vets, we risk not being trusted and licensed in future (or other people being appointed instead).

*

We might think we should follow others' instructions when they have a legitimate moral authority over us (within their areas of authority), even when that involves doing what we would otherwise think morally wrong (and we have been unable to influence those instructions). We might think this will improve consistency or public trust. However, we are still responsible for our decisions, our behaviour and our outcomes, and we should not simply 'pass the buck' to bosses or regulators. Indeed, this gives us the worst of both worlds: we remain accountable while relinquishing control. We might also question whether we really should want the public to be able to 'trust' us to follow instructions that we think are wrong.

Alternatively, we might think we should just do whatever we would otherwise have thought right, simply because it is right. However, this approach could lead to worse outcomes (e.g. if colleagues no longer trust us, or regulators remove our ability to help animals in future), and would arguably suggest that other vets should also ignore any rules they think are wrong (but which we think are right). As another alternative, we might think we should factor the existence of the instruction into our decision-making, considering the scientific and moral expertise of the regulator and the potential outcomes of non-compliance (e.g. the risk we lose our licence to practise). This would suggest we should usually comply, but not in exceptional cases where the outcomes of compliance outweigh those of noncompliance.

Ideas

 Legally bare

Sometimes we treat our laws or professional rules as if they answer every ethical question, and avoid moral questions on how and whether to follow them or to change them, or what to do where the law is silent.

 Being professional

Being a professional suggests competence, integrity, selflessness, trustworthiness, respectfulness, reliability and sound judgement. However, the term is rather vague and can defend or criticize almost anything. We should try to avoid ever thinking we are being unprofessional, but we should not use the term without being clear precisely what we mean.

 Comparing ourselves with others

We often assess our ethics in comparison with others, defending our behaviour as common, normal, popular, traditional or part of our culture (e.g. slavery, blood sports, animal testing or pollution). It certainly feels more comfortable to be morally 'within normal range'. However, this confuses 'common' and 'right'. We should instead evaluate even (or especially) common views.

 Passing the buck

We often do not want to disappoint, embarrass or upset people, and fear being challenged or reprimanded. It can also feel easier at the time to let others make our decisions for us. Afterwards, we might then feel we were weak-willed, unfairly pressured or taken advantage of. We should instead treat such instructions as factors to consider.

Reflections

- Are we taking others' views adequately into account (to challenge our own), while also being sufficiently independent thinkers?
- If we are worried what others might do:
  - (When and how) can, and should, we help one another to improve clinically and ethically (e.g. by discussions, policy-making or enforcement)?
  - Are we basing our assessments on accurate information (e.g. will the client actually go to another vet and would they provide the management we have refused)? Can we change these (e.g. by speaking to that vet)? Do they only provide the management because they think that, otherwise, we would?

- Is there a legitimate reason for us to do something rather than someone else, or vice versa (e.g. specialism, reducing a patient's transportation or owner's cost)? Are we biased by our own interests (e.g. profit) or wanting to avoid a difficult case?
- Are we subconsciously trying to 'pass the buck' to others to avoid feeling responsible?
- For instructions:
  - Does whoever gives us an instruction have the moral expertise, authority and permission to do so? Does that apply to us, to this case, and to every option? Can they morally mandate or only prohibit?
  - Is the instruction a moral one for them to make (to ensure ethical behaviour or to improve consistency)? Have we undertaken to follow this instruction? Was that undertaking limited only to behaviour that we consider ethical? Is the wrongness of failing to fulfil that undertaking outweighed by the wrongness of the behaviour? If not, should we ignore or even oppose it (and what would be the outcomes of doing so)?
  - Is the instruction a request, a suggestion or an order? Are we being told to use our professional judgement or to follow blindly?
  - Is the instruction one that should always be followed or are there legitimate exceptions (e.g. to follow another instruction)? If so, what are they? Is this case one of them?
  - What would be the actual outcomes if we disobeyed this particular instruction? What would be the outcomes if everyone disobeyed similar instructions when they thought them immoral?
  - Can we change the instruction, e.g. by speaking to whoever gave it? If so, should we delay following it in the interim?

4.3   Our relationships with all animals and people

So far we have considered our relationships with specific individuals. We also have relationships with, and within, whole populations. More widely, we might feel we have relationships with all 'animals' and 'society', who are themselves interrelated within societies and ecosystems.

Let us recall cases that concerned whether we did what society expects us to do.

- What are our responsibilities to human and animal society?
- Do we, as vets, have particular responsibilities to animals who are not our patients (that other people do not have)? If so, what?
- Should we be morally concerned with societal progress, knowledge or economic growth?
- What are our responsibilities to nature as a whole? Should we try to ensure animals are as 'natural' as possible? Should we give greater protection to endangered species than to commoner species, or simply consider what is good for each individual animal?
- Should we be particularly concerned with protecting the human species (above its individual members)?

In modern times, environmental issues present a major ethical concern. We might think we should make efforts to prevent climate change or environmental degradation, both through our direct actions (e.g. our practice's carbon footprint) and what practices we help support (e.g. polluting farming practices). Indeed, we might think we should do all we can to minimize our environmental impact. However, we might think such an approach would be unfair insofar as other businesses undergo more modest limits (especially if we think veterinary work inevitably has a negative impact, given the resources and waste involved and the environmental impacts of our pet and farming clients and patients).

More widely, we might believe nature has an inherent value, and that animals' lives should be as natural as possible (e.g. their genes, environments and behaviour). However, we might find it hard to apply ideas of naturalness to farming, competitions, artificial selection, pedigrees, research, domestic species and veterinary treatments. Alternatively, we might believe nature is important only because its loss impacts animals and people (e.g. ecosystem degradation and captive conditions cause animal suffering, and species loss is a symptom of many animals' unfitness or deaths). Additionally, we might reckon that natural lives involve suffering for individuals (especially if we release tame, domesticated animals into the wild), which human intervention can avoid, so we might think we should promote natural states and processes when only they are better than artificial ones, but not when they are actually worse for the animals involved.

Assuming we cannot unilaterally stop human environmental impacts, we might think we have a role in the conservation of endangered species (and perhaps 'unnatural' breeds or bloodlines). However, we might think this approach is unfair, since it suggests treating individuals differently depending on the prevalence of their species (I do not consider myself any less valuable because humans are not endangered). We might also argue that saving the last animals just tackles the symptoms, and our desire to 'keep' a species or breed should not override our responsibilities to individuals (e.g. we should still euthanize suffering endangered animals in breeding programmes, and not support breeds whose members suffer inherited pathologies). Alternatively, we might simply see species and breeds as fluid taxonomic labels, ignoring species in how we treat individuals and instead trying generally to prevent the causes of Anthropocene ecological catastrophes.

*

We might feel particularly morally concerned about saving humanity (although humans are generally neither vulnerable nor beneficial parts of natural ecosystems), either because we believe humanity is uniquely special or because we think anyone should try to save members of their own species (but not of their genus, family, order, class, phylum or kingdom). However, we might recognize this does not mean every human concern (e.g. profit) is more important than any animal's outcomes. Alternatively, we might think a human bias is an unjustifiable prejudice, or that trying to save humankind and animals is not mutually exclusive but, rather, part of the same concern for the planet's life as one diverse and interconnected ecosystem.

We might also value some idea of 'progress' (e.g. knowledge or economics). However, we might not think a label of 'progress' justifies any and every informational, economic, technological or genetic development – any 'progress' needs to be towards a legitimate destination and using acceptable methods. So we might reckon knowledge is desirable only when research is humane and the results benefit patients. Similarly, we might think we should aim for economic growth for our practice, clients or country only when it reflects genuinely better lives. (And we cannot then define better lives economically, since monetary measures are unreliable measures of real value, and may ignore ethical concerns such as fairness or impacts on children, animals and nature). We might think we should morally ignore any benefits that accrue from immoral behaviour (e.g. data from inhumane research or profit from exploitative practices).

*

Society also has relationships with animals, in particular how its cultural and legal structures affect how they are treated by humans (e.g. animal industries), and how humans affect their environments (e.g. land use). We might think social structures can reinforce or embed norms that help animals and others (e.g. through animal protection, land use and consumer rights legislation). We might think they can redress imbalances in outcomes (e.g. compensating owners after 'negligence'). We might worry that social structures also create or embed structures that mean innocuous behaviours by individual members combine to some overall wrongdoing (e.g. multiple agents in a supply chain), or reinforce harmful attitudes or relationships (e.g. prioritizing property rights over responsibilities).

We might think we have more responsibilities to all animals than other people have. We have greater knowledge of animals and, as a profession, we have claimed to know and care about animals. If we then appear to ignore, tolerate or defend animal management practices, then our silence or support could be taken to mean that we consider those practices to be acceptable or even beneficial for animals. For us, inaction is not a neutral option. Alternatively, we might think we have no particular moral responsibility for animals who are not our patients (any more than anyone else). However, this would suggest we should represent ourselves as mere technicians and allow others who do take on this responsibility (e.g. animal welfare charities) to have a greater impact on animals than we do. We might not like that option. Nonetheless, we cannot have it both ways – we should not claim a responsibility but not fulfil it.

Ideas

 Animal advocacy
We might think that we, as veterinary professionals, should be advocates for what is right for animals, in all contexts.

 Endemic wrongdoing
Some behaviours may be so prevalent in our society, or integrated into its structures, that they affect everyone, even those of us who would change them if we could.

continued

Maintaining compassion

Our compassion for animals may be why many of us became veterinary surgeons or nurses. However, we can sometimes lose our focus on this, through the many other pressures that we face in our work and because of our limited energy and time. We should try to maintain our original motivation, feeling rewarded by our impact even in the face of challenging clients and situations.

 Overwhelmed

We can also feel overwhelmed by the amount of animal suffering in the world, and powerless as an individual vet. We should recognize that we cannot help every animal everywhere, but still value the help we do provide.

 Robotic medicine

We might not 'feel' any emotion towards society (in comparison with the very real pressures from individual people). Indeed, thinking and acting with professionalism, objectivity and integrity might feel particularly 'cold' and dispassionate. Instead, we might try to think of our professional approaches not as the absence of any compassion, but as an impartial recognition of what matters to *all* animals and people.

Reflections

- Is there a clear legal mandate or prohibition imposed on us? Could we avoid the situation in which that law applies (e.g. not admitting non-indigenous species except for euthanasia)?
- Have we given the impression that we, as a profession and as individuals, particularly care about animals? (How well) are we fulfilling or betraying that undertaking? How could this affect our ability to help animals in the future? If we are not, should we change what we do, or the public's expectation?
- What societal expectations and undertakings should we fulfil? Should our professional responsibilities ever prevent us from doing what we feel is morally right?
- Should we want societal 'change' even if it does not lead to greater benefits overall? Is financial performance (of a country, farm or practice) a measure of improvements in the lives of the people and animals involved?
- Is something necessarily better if it is more natural? Or is it only better than some unnatural states? Should we ensure domestic and captive animals' outcomes are at least as good as they would be naturally (i.e. in natural environments, with natural genetics and ontogeny)?
- Should we make clinical decisions based on what is best for animals as individuals, regardless of species or breed (except insofar as these are practically important, e.g. in signalment)? Should we ever harm an individual animal to try to maintain its breed or species? Would we be better trying to prevent the root causes (e.g. habitat loss or pollution)? Are breeds created by humans intrinsically worth perpetuating (or only for some humans' enjoyment or income), even if that causes members to suffer?
- Should we consider humans as particularly important? If so, how and when?

Part B: Developing Ethical Skills

We have considered what ethical views we might hold. Applying our ethical views to practice takes skill. Before considering how our ethical views can apply to practice, we can consider what key skills we can develop. Ethics can help us ensure we develop in ways that build up our 'ethical muscles' (to which Julius Wolff's law on bone remodelling in response to loading applies as much as anything else), improving our resilience, confidence, behaviour and effectiveness.

Understanding Situations

5

5.1 Ethical investigations

One important ethical skill is identifying what matters morally in each case presented to us. We can think of cases as having particular features (like antigens) with which our views engage (like antibodies). We may get better at predicting why cases might be challenging or contentious and seeing what concerns and views could be useful in making our decisions. We may find it particularly important to be able to see when issues have multiple ethical aspects, to help us ensure we make decisions based on a comprehensive range of factors and ethically relevant principles, and to ensure we can have meaningful ethical discussions with other people involved. So the key skill here is to identify *all* the moral 'aspects' of a case.

We might start by sketching out the 'anatomy' of all relevant general concerns that might apply. Our ethical investigatory skills might be considered more developed when we can consider a wider range of issues, across many circumstances, and when our concern relates to people and animals who are increasingly different to us (e.g. across gender, race, age and species). As we get more practised in recognizing, and understanding, moral concerns, we can also see how different ethical concerns connect with one another (e.g. whether one underpins, overlaps, or conflicts with another). Discussion with others is one way to improve our range.

In our investigations, we should avoid – or at least minimize – predisposing ourselves towards or against particular moral decisions. We should avoid words that frame things negatively (e.g. 'pest', 'vice' or 'sacrifice') or euphemistically (e.g. 'put to sleep'). Even technical words can elicit or avoid emotions (e.g. 'slaughter', 'destroy', 'cull' and 'euthanasia'), and the academic tactic of trying to create new words (e.g. 'devitalize') may appear a deliberate attempt to avoid certain implications. We should try to see past our descriptions, by considering not only how our word choices affect our ethical viewpoints (and those of others) but also what they reveal about our own assumptions.

Ideas

 Needing measurement

Some important ethical concerns do not relate to measurable parameters (not everything that matters can be measured). Even economic measures of preferences miss many values for many people and animals.

 Partial reflections

Sometimes we only consider part of the issue when making decisions. For example, we might consider only one set of people affected. We should instead consider everyone, and every relevant factor.

5.2   Evidence-based ethics

We can think about ethics in the abstract, but we also need facts about our cases to make appropriate decisions in practice (e.g. facts about our patients, clients, finances and the law). However, facts alone cannot tell us what to do, and we also need to identify our ethical concerns that make facts salient for our judgement. We can think of facts as being ethically indicative or contraindicative of particular options (e.g. the prediction of pathological pain might indicate treatment; predictions of iatrogenic side-effects might contraindicate it). Indicative factors suggest we should perform some behaviour. Contraindicators suggest we should *not* perform a given behaviour (and they cannot suggest a behaviour on their own – except indirectly by ruling out one behaviour but not others).

We might want as many facts as we can find. However, too many irrelevant facts can make our decision-making sluggish (somewhat like amyloid build-up can reduce organ function). So we need to learn to distinguish which facts are relevant and which are irrelevant. In any case, relevant facts are often unavailable (e.g. on initial presentation), unclear (e.g. ambiguous presentations), uncertain (e.g. all scientific facts represent statistical associations), unreliable (e.g. tests with potential false positives), unmeasurable (e.g. status of cellular immunity) or too complex to analyse (e.g. metabolomics). Nonetheless, we often have to make a decision or risk delaying the right behaviour. So it is important that we are not only skilled at ascertaining and applying facts, but also discerning but also discerning which facts we actually need to obtain, in using them wisely and in accepting that we rarely have a complete information set.

One approach to ascertaining relevant facts is by screening available facts against our ethical views that might be relevant, to understand how they might affect our moral decisions. Alternatively, we might start by identifying pertinent ethical views and then identifying relevant facts, by generating conditional 'hypotheses' and 'research questions' (e.g. if we believe we should protect endangered species, we might want to identify an animal's International Union for Conservation of Nature classification).

In practice, we might combine repeated, iterative steps of establishing facts, identifying relevant ethical views and generating hypotheses. Often, for common or simple cases, some steps might be subconscious, because it seems obvious what facts or views are relevant (e.g. that an injury is painful). It is very much like – and is part of – clinical history-taking and examination.

Ideas

 Statistical confidence
Having more facts or certainty can make us (and owners) feel more comfortable and confident. However, we should deal with uncertainty intelligently, gaining more information when useful rather than as a crutch so we feel better.

 'It's only opinion'
We sometimes talk of ethics as 'having no right answer' (in comparison with science). However, most of us would defend at least some of our views as right (e.g. that rape and animal abuse are wrong). We should accept that (as for scientific theories) we might change our minds, but this does not mean we should dismiss it entirely.

 Facts alone
We might feel more comfortable considering facts (e.g. clinical data or relevant laws) than ethical views. However, facts alone cannot lead to decisions on what to do – we also need moral views to combine with facts so that we can make decisions.

5.3   Ethics-based evidence

We should also consider the ethics of obtaining information. We should seek data that are available in a timely manner. We should not unduly avoid or delay making decisions when facts are unavailable. We should avoid obtaining data in ways that involve harms (e.g. anaesthetic risks, and financial costs) that are not justified by the increased insight they would provide. In conditions of uncertainty, we still need to be able to make decisions using the data and judgements available. We might cautiously avoid what seems the worse error (e.g. providing analgesia when unsure if the animal will be in pain). At other times, we might use our subjective 'gut-feeling' or confidence in what is right.

We might believe different facts, depending on our expertise, experiences and our prior ethical views, which can predispose us to believe certain things (e.g. those who think we should have no responsibilities to animals might dismiss evidence of non-human cognitive capacities). Furthermore, many facts require prediction, empathy, judgement and evaluation (e.g. clinical prognoses). Consequently, people might disagree about the facts of any case, even if presented with the same data. This means that, far from being the most reliable and agreed aspects of a debate, facts can sometimes be the most contentious. It also means that another ethical skill is to be open to facts that do not support our ethical views.

Ideas

 Feeling for others
One key moral ability is identifying and understanding others' feelings in a way that involves us experiencing parallel emotions too.

 Ethics of belief
Our acceptance or scepticism of facts should itself be ethically right. For example, we might think that we should be as open to the possibility of others suffering as we would want them to be open to our own suffering.

5.4   Listing options

An important skill, in order to work out what we should do next in any case, is to identify what we *could* do. This then frames our ethical decision as a choice between available options.

We might find it useful to start by recognizing all possible options – even those that seem unattractive initially – to avoid prematurely discounting an option that we find is actually the least bad. Conversely, what initially seemed like the obvious best option may sometimes get ruled out later by our ethical reasoning, so we then need to turn to suboptimal options to make the best of a bad situation. Furthermore, ruling options out right at the start often presupposes a particular ethical viewpoint that these are always wrong. So we should avoid predetermining our reasoning unless we are absolutely sure it is right. Ideally, we should create a list of all options that avoids any overlap or omissions, from which to choose (i.e. the options are mutually exclusive and collectively exhaustive).

We can then rule out impossible options (e.g. treating incurable diseases), because we cannot be morally obliged to achieve the unachievable. We might also rule out options that we think are always completely morally unacceptable in every possible case (if we think any are). By ruling out such options, we focus our decision on genuine options. For example, if medicating an aggressive animal is effectively impossible, then euthanasia may be the best option within the constraints of reality (e.g. the owner's unwillingness to keep it and the inability of charities to rehome such animals).

Ideas

 Mission impossible
Sometimes we feel bad for not being able to help where we realistically cannot. We should instead only feel we have duties to do what is possible.

 Presumptive exclusions
Sometimes we rule out options too early (e.g. that closing a farm would be unfeasible or impractical) before we consider them properly. We should instead only rule out genuinely impossible options and morally consider all others.

 Multiple choice
Sometimes we try to choose between options when we could choose both (e.g. supporting industry *and* animals; remain compassionate and professional; love animals and euthanize them) or neither of them (e.g. allowing neither tail-docking nor tail-biting; laboratory research on neither rats nor children). We should instead consider choosing none, some or all options.

5.5   Secondary options

Sometimes we decide on options in terms of outcomes (e.g. cure versus palliation versus euthanasia). In such cases, we have a second decision to make as to how to achieve that outcome (e.g. we might treat an abscess medically or surgically). This is especially important when considering less obvious alternatives that would tackle the root causes of a problem (e.g. outcrossing a breed as an alternative to genetic testing to achieve healthy animals). So we should not make assumptions about what is unchangeable.

When we have decided on a particular option (e.g. what surgery to provide), we are often faced with secondary decisions (e.g. whether an owner will give permission, what analgesia to use, what to charge). We might make these after making the primary decision, but it is sometimes useful to make secondary decisions beforehand, to inform our primary decision. For example, to decide whether to perform a surgical intervention, we might need to pre-assess what analgesia we will provide (and not assume we will control the perioperative pain perfectly). The secondary decisions can then serve as conditions on which our primary decision was made. However, we should ensure that provisionally making secondary decisions does not make us assume we have made the primary one (e.g. focusing on how to refine a surgery or animal experiments presupposes that they are performed).

Ideas

 Conditions

We might decide on an option and then decide details. But often our choice of what to do depends on those details (e.g. that an owner will give permission). If so, when we cannot fulfil those conditions, we should revisit our original decision.

 Clinical necessity

We occasionally think something is 'necessary' or 'inevitable' in a sense of unavoidable. But more often, something is necessary for something else (e.g. for an animal's or business's survival) or inevitable because of previous choices (e.g. how society is set up). So we should never just accept something as 'necessary', end of story, but always clarify for whom, for what, when and why.

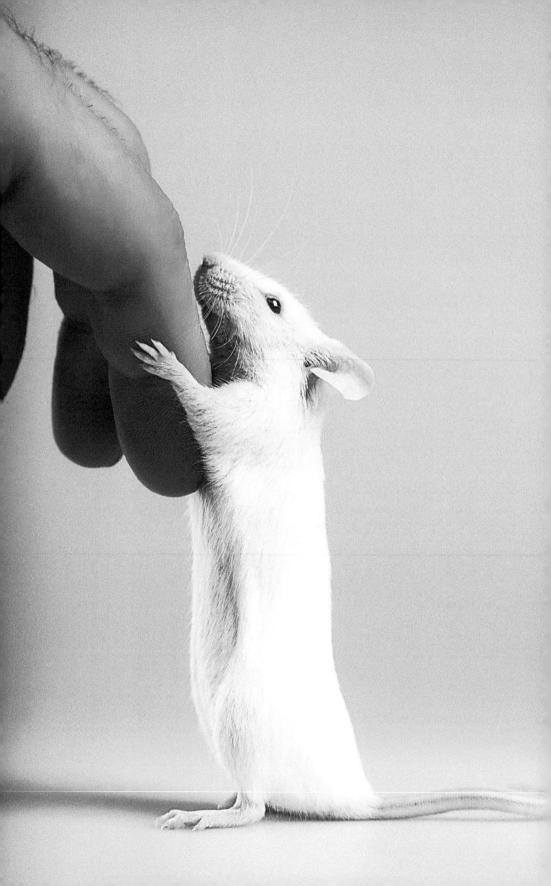

Understanding Others

6

6.1 Recognizing moral patients

As clinicians, we have recognizable clinical patients that we aim to help directly in our work. As vets, we might also morally think of our 'patients' in a wider sense of everyone to whom we should behave morally. So, alongside the skill of determining *how* we should act morally, we should also try to get better at determining *to whom*. Key to both questions is being able to recognize *why* we should act in those ways to them.

In trying to answer these questions we need to be able to identify views that are sufficiently generic to recognize all potential patients (including ourselves from others' perspectives) and relevant effects (which are important to others, if not to ourselves). Such generic reasons cannot be completely personal. I may be a male, but I cannot say we should be kind only to males morally since the generic equivalent for a female (i.e. to treat only females morally) would leave me out. I may like poetry and chocolate, but these seem unsuitable for dogs. We need an approach that we can use for different species, genders and individuals, while still taking into account variations between them.

We can also try to identify particular traits of animals that mean we should treat them morally. For example, we might believe we have moral duties to animals who can experience pain and pleasure (which includes most of our patients and many others). We might think we have moral duties only to animals who can be moral to others (which excludes most animals and human babies). Once we have such categories defined, there is then the challenge of identifying which animals have those features (i.e. which animals can experience pain or demonstrate morality). This is an exercise in which we might want to err on the side of caution, in the same way we would want others to err on the side of caution for us.

Ideas

 Do-as-you-would-be-done-by
One consistent approach is to treat others as we would want them to
treat us, or (much more demandingly) to love one's neighbour as oneself.

 Only we count
Some approaches to ethics assume that our moral responsibilities
to humans are always more important than those to other species.
However, there are no phenotypic traits that all and only humans have
to justify this. So we might think this is essentially just an emotional
predisposition.

 Limited claims
We might think of responsibilities as moral claims made on us. However,
this risks biasing us towards people who can say what they want, above
those who do not or cannot (e.g. children, animals or selfless people).
We should instead ensure we identify and consider all moral obligations.

6.2   Ethics discussions

Perhaps the hardest skill is having meaningful ethical discussions. It can be
easy to explore only a narrow and unchallenging range of ideas and options
with people with whom we agree. Conversely, it can be easy to talk past
or be confrontational with those with whom one disagrees, in ways that
sounds convincing to oneself but miss the point for others. Indeed, it can be
easy not to have moral conversations at all, especially if we are not confident
in sharing our views or accepting challenge.

We can share our views so that others can challenge or help reinforce
them and so we can learn from one another. Ethical ideas can infect other
people, spreading like genes or pathogens. In some cases, we might try to
avoid taking in others' viewpoints when they seem morally 'pathological' (i.e.
harmful to our moral functioning). However, excessive ethical sterility risks
us being overly naïve and less able to resist ethical challenge when it does
come. We also want to gain beneficial ideas from others, to give us our own
'commensal' viewpoints to help protect us from pathological ethical germs.

For this, we should also develop the skills of having quality ethical conversations. We should be open to exploring ideas together, openly, meaningfully and genuinely, and with an expectation of changing our own views.
Debate can challenge or reinforce personal views; discussion can identify
common ground in a group and refine those shared views to a joint conclusion (rather like soft tissue surgery; Fig. 6.1).

Fig. 6.1. Surgical analogy for ethical discussions.

Ideas

 Talking to win

We sometimes use rhetorical devices that help us feel we have won an argument, such as implying our moral superiority (e.g. 'I am trying to cure cancer/love animals'); giving pejorative labels (e.g. 'emotional', 'bunny-hugging' or 'animal rights'); or shifting our point mid-argument. We should instead explain and explore our views openly and critically – actively hoping to change *our* mind too.

 Focusing on disagreements

Sometimes, people with different ethical views may actually agree on many conclusions (e.g. veganism might be based on concern for suffering, death or environmental protection). We should recognize our practical agreements without being too distracted by philosophical differences.

6.3   Understanding others

We should also try to understand others' viewpoints, so that we can engage with them meaningfully. We can understand others in multiple ways. We can understand them ethologically, in terms of the biological and psychological causes of their behaviour (e.g. we might understand aggressive owners in terms of their emotions, brain function, stimulatory cues, previous conditioning, upbringing, or partial evolution). But we can also try to understand their ethics from a more philosophical and ideological perspective.

We might also want to consider how people want to make decisions. Some people might want to ignore probability or even avoid making treatment decisions at all (e.g. if they feel these demonstrate their devotion to their pet). Some people may want us to make their decision for them, or at least to give them clear professional advice (rather than mere information). Some people are concerned about their moral traits; others about their own intentions or behaviour; others about their relationships; and others about the outcomes of each option or some wider concept (e.g. nature) (Fig. 6.2).

More generally still, we might consider whether people make decisions by using their moral intuition or emotions directly; or copying others; or trying to live up to some ideal; or trying to achieve something. Of course, we can also consider whether people are making decisions based on their moral views or

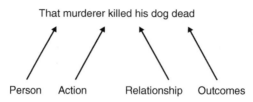

That murderer killed his dog dead

Person   Action   Relationship   Outcomes

Fig. 6.2. Sentence to illustrate key ethical foci.

their self-interest; and whether they are making decisions themselves or being pressured by others. However, we should avoid assuming others are being selfish – people are usually motivated by something they think is morally worthwhile, or at least have some self-justification for their behaviour, even if we cannot see it immediately, or consider it insufficient, erroneous or outweighed.

We also sometimes fall into a trap of thinking we understand (and perhaps share) others' views, because we are understanding similar words in different ways. Many ethical words are somewhat ambiguous (e.g. 'respect', 'reasonable', 'appropriate', 'objectionable', or 'in accordance with the highest ethical principles'). Sometimes people might fail to articulate their assumptions (e.g. 'It is unacceptable' without specifying for whom) and we might mistakenly read views into others' words that they did not actually express (e.g. 'We should protect laboratory animals' as implying 'We oppose research'). In some cases, people will deliberately use ambiguous words to give the impression of an ethical view or to prevent us recognizing their real motivations. So there is a skill in spotting these fudges and platitudes, and clarifying what is meant or missing, or reserving judgement until we are more convinced.

This is important when we want to influence other people, to help us avoid talking to them in ways that would persuade us but are unconvincing to them. Once we understand, we can highlight facts they will consider relevant. For example, some owners might consider that what is natural is more important than what reduces suffering, or be more concerned about the suffering they cause than with the suffering they merely allow. It is also important for us to compare our views, and see if we need to improve them.

Ideas

 Echo chambers

There is a danger of having conversations within mutually reinforcing echo chambers. This means we miss useful challenges to our views and reinforce them, fail to understand others' views, cannot form shared views, and potentially increase separations.

 Non-specific diagnoses

People sometimes talk in ways that leave the main moral questions unanswered. We should instead always try to explore (and explain) what is meant and what is missing.

6.4 Evaluating others' ethics

Our diagnostics can also be applied to judgements of behaviour. Insofar as ethical views imply that particular views, decisions, behaviours and statements are right or wrong, it is perhaps inevitable – and right – that we judge others' behaviour.

However, we should also ensure we are making correct judgements and in morally acceptable ways. We should judge fairly, sympathetically and consistently, and we should not be harsher on others than on ourselves (or vice versa). This is made difficult because we can always identify the goals and factors that justify our decisions or excuse our weaknesses, while missing others' different moral aims, viewpoints and contexts. We should therefore try to understand others' viewpoints and situations – '*tout comprendre, c'est tout pardonner*'.

We might also think we should take into account contextual facts that could affect others' views, for example their culture or upbringing. We might think these are relevant and legitimate determinants of their view, i.e. it is acceptable (or even right) that different people have differing ethical views depending on their context. Alternatively, we might think that ethical views should be consistent and universal, and that such contextual factors might psychologically explain their views, but cannot philosophically justify them. In either case, we might reflect on whether we disagree because of our unjustifiable cultural contextual predispositions.

Ideas

 Tilting at windmills
Sometimes our dismissal of ethical viewpoints might be based on misleading or oversimplified misconceptions of them (e.g. animal rights as affording animal suffrage). Instead, we should try to understand other viewpoints before dismissing them and, if we think someone else's viewpoint is wrong, we should consider whether we might actually have misunderstood it.

 Excusing and forgiving
We might excuse others' failings because of contextual factors (e.g. pressure from others; impaired decision-making due to grief or stress; or because they are themselves victims of abuse). We might forgive others where, even if we think their behaviour is wrong, we refrain from (or remit) our judgement of them as a person.

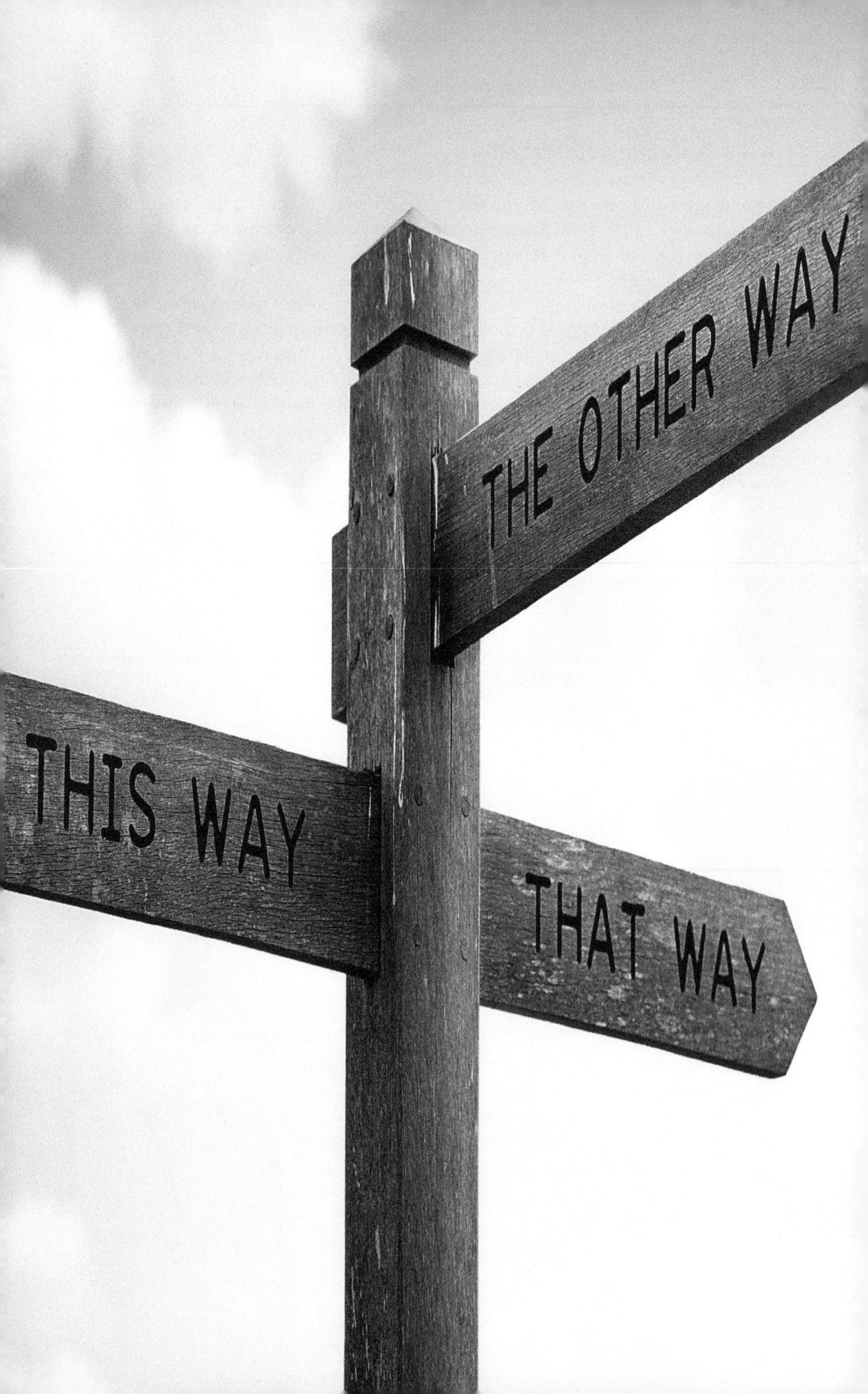
THE OTHER WAY
THIS WAY
THAT WAY

Approaching Decisions

7

7.1 Emotional intelligence

We can also learn to use our emotions in our assessments and decision-making. We might see concern, pity, fear, indignation, guilt and hope as relevant indicators of ethical factors, which we should take into account (at least as hypotheses to prompt further reflection). We might try to ignore our emotions and make decisions purely on logical grounds, but this seems misguided – and probably impossible – given that our moral viewpoints ultimately link to emotional evaluations. Our ethical views are sometimes (and perhaps always partly) based on emotional foundations (e.g. it is hard to say why pain is morally bad; it just feels bad to experience or see it).

In particular, we might think that, by doing what will avoid us feeling guilty, we will do what is right. However, this approach can be used the wrong way round – or at least be circular – insofar as we should generally feel pleased or guilty depending on whether we think what we have done is right. Also, avoiding guilt or regret may psychologically push us towards passing the buck, deferring decisions, doing nothing and never reviewing previous cases (and therefore never improving our ethics). Conversely, it is wrong to assume something was acceptable because we do not feel guilty, as guilt has poor sensitivity and poor specificity as a measure of wrongness (e.g. sometimes we feel guilty for doing our best to mitigate owners' faults). Nevertheless, we may use guilt as an indicator that something is at least ethically questionable.

There is also a wider danger that we try to avoid feeling emotions, perhaps because they are unpleasant. For some cases, we might use internal coping mechanisms that bury the problem deep down (although it may resurface later on, sometimes more aggressively), or we might feel it is socially unacceptable to display emotions, or we might find that constant exposure to difficult cases scleroses our affective processing. For other cases, we might actually think it beneficial to 'harden our heart' in order to implement what we have decided is right, despite it being unpleasant (e.g. euthanizing a suffering but beloved pet). Even in these cases we should recognize, accept and reflect on our emotions.

However, we do need to learn when our 'gut feelings' are reliable and when further reflection might improve our decisions. We might sometimes 'emotionally misdiagnose', for example mistaking pity for guilt (while doing something wrong makes us feel bad, that does not mean we must have done something wrong whenever we feel bad). Similarly, we need to learn to differentiate our genuine moral emotional responses from our self-interested emotional responses (e.g. approval feels somewhat like gratitude; indignation feels rather like anger). This differentiation can be particularly hard when we feel we are victims of others' immoral behaviour. So we do need to analyse our emotions carefully and critically (even, or especially, when they are strong).

Ideas

 Over-caring
We sometimes see so many cases of suffering that we down-regulate our (conscious) emotional responses. It can be a symptom and source of stress. It can feel like moral exhaustion, withdrawal, hopelessness or stress.

 Inner voices
We sometimes feel like we have an internal faculty that critically judges behaviour, and ourselves, against internalized standards.

 Fear of failure
Sometimes we fear failure so much that we continue treatment, avoid euthanasia, or avoid addressing difficult situations at all. Instead, we should ignore our pride and do what we feel right, even if we could do 'more'.

7.2   Spiritual guidance

For many people, our ethical views are also based on more spiritual considerations of the world that transcend our worldly concerns. Sophisticated ethical concerns are a key part of all major world religions and religious moral views are legitimate, meaningful and authoritative for many people. Our individual and shared faiths often underline or reinforce ethical views and traits, such as compassion, charity and respect for nature. They might also provide additional moral concerns (e.g. considering particular species as sacred).

Many religious people championed animal protection or compassion, often in combinations with a concern for humans (e.g. St Francis, William Wilberforce, Albert Schweizer), and many religions believe animals have souls, comparable to, and moving to and from, humans (e.g. Hinduism and Sikhism). Some, such as Jainism, have very strong approaches to animal protection. We might be inspired by sacred scriptures and enlightened by their interpretations of previous and recent scholars. We might also try to infer ethical views from facts by identifying a purpose or end underlying the world. For example, we might think that the purpose of animals is to hunt, sleep, purr, etc. (or to be used by humans), or that the purpose of humans is to live socially (or to be used by other humans).

Alternatively, we might think that faith is somehow separate from modern secular ethics (sometimes seeming like agonistic/antagonistic muscle sets). In fact, many of our seemingly secular cultural, ethical and legal norms can be largely traced back to, are implicitly informed, imbued and enlightened by, and still correspond with, religious views (whether we realize it or not). Indeed, we might think these ideas make less sense, or are harder to defend, without their theological basis (e.g. sanctity of life or human dominion without the idea of being in God's image). So, whether or not we can feel religious ourselves, we should still try to understand others' religious ethics.

Ideas

 Fitting with the world
We might think we should live in accordance with, or fulfilment of, the underlying laws of the universe.

 Over-secularization
We sometimes dismiss religious views as not being objective. However, our criticisms are often based on unrepresentative caricatures of believers' views, and many non-religious views are also based on non-scientific 'faith' (e.g. atheism, evolutionary progress or human improvement). We should instead take religious views seriously – and be open to the idea that they may be right.

 The wings of a dove
Death, nature and God can cause strong existential, spiritual and numinous emotions, both positive (e.g. awe and love) and negative (e.g. fear and angst). We should use these emotions when they are powerful and positive, perhaps turning to spiritual or philosophical guidance to address their roots, but be suspicious when they suggest harming animals.

7.3 Moralogic

We might think we should be logical in our ethical thinking. We might think it will help us to make decisions, to be consistent, to persuade other people, and to avoid accusations of hypocrisy, inconstancy or irrationality. We might also feel uncomfortable when we have inconsistent or logically conflicting beliefs. Setting out our logic can help us highlight our assumptions, identify errors and predispositions, and explain our views to others. However, it is surprisingly difficult to argue why we should be logical without using logic (which makes it somewhat circular).

We might even think being moral is fundamentally a matter of being logical, as opposed to the emotional and instinctive behaviour found in all animals. We might think illogical behaviour (e.g. making false promises) is itself immoral, although we might struggle to think of many examples of inherently illogical behaviour (we might even think of examples where it seems logical to make false promises). However, focusing solely on logic, and not emotion, makes it difficult to see why our pre-logical emotions are morally important (e.g. avoiding pain). We might also worry that logic can be misused (e.g. to justify predetermined views) and is inefficient (e.g. when we need quick decisions in practice). So we might think we should make decisions by combining logic and emotions.

Broadly speaking, statements can be stated (e.g. sentences using 'is'/'is not' or 'should'/'may not' or 'may'/'should not' – noting that the negation of should is not 'should not'). They can then be combined (e.g. using 'and', 'or', 'therefore', 'because', 'if', 'unless'). Different statements can be compared and considered equivalent or that one is included within another (as if phagocytosed). Multiple statements can be combined into longer chains, and these chains might then be reduced down once we know the middle terms are redundant (like anastomosing two intestinal loops). Logic suggests ways in which statements can be validly combined or refined.

We can use logic to progress from our fundamental starting points (our basic emotional evaluations and motivations, and our factual beliefs) to implied conclusions. We might analogize ethics to engineering. We try to start from solid foundations and use quality raw materials. We want to create a strong structure to support us in our lives and on our journeys. For a house, we start from the bottom and build up. For a bridge, we construct between solid points. For a pier, we might start where we are and build into the unknown.

We might aim to be consistent in our views and behaviour, treating similar facts similarly and applying values and logic in consistent ways. We might even aim to have a perfectly complete set of coherent moral beliefs. However, this seems unrealistic in practice – many of us have myriad conflicting ethical beliefs across different cases. For example, we might believe that we should aim for the best outcomes for our patients, while also wanting to avoid crossing particular lines such as stealing from clients. Perhaps, given the many influences on our moral development, it is inevitable and psychologically understandable that we often have multiple, potentially incompatible views (and often find the implications of single views unpalatable).

Ideas

 (Tauto)logic

Logic is essentially about following formal rules from starting points to conclusions. Pure logic can effectively be reduced to a series of sentences that mean the same thing. So, to be useful, it needs to be combined with true facts.

7.4 Spotting logical lesions

One of the key skills in ethics is to spot where our ethical thinking is imperfect. Identifying errors helps us see where supports are weak – where we either need to find better supports or move out of (and perhaps demolish) the construct as unsafe. Spotting such weaknesses is an important skill for checking our own and others' reasoning. Even if we do not believe our ethical views need to be based on logic, we might still think we should not be visibly illogical or falsely claim to be logical. So spotting contradictions is important.

Medically, we might think of these errors as lesions, damaging the integrity of the tissue either by missing steps (like fractures), reaching undue conclusions (like dysplasia) or getting out of control (like neoplasia). Some of these can be benign, but some can displace other logic (like space-occupying lesions), cause functional problems, or spread to other ideas. Architecturally, to stretch the above logical analogy further, we might say that some constructs are mere 'facades' (like Hollywood set designs) that look plausible but are really poorly supported.

One type of weakness is where we actually disagree with ourselves. Sometimes we might contradict ourselves logically (e.g. if an owner asks for our opinion but we refuse to provide it out of respect for their decision-making). At other times, we might have factual conflicts (e.g. if we think that we should always make the prevention of suffering our main priority and always get owner permission for any treatment, when we factually cannot always do both). Note that a single ethical view cannot be unreasonable or illogical in itself – only combinations of views can be incompatible or inconsistent with each other. In ourselves, we should try to be consistent, and ideally all our views should fit perfectly together (like healthy bodies do).

Another weakness is when our factual assumptions are actually wrong. It can be useful to think of arguments that would be right if certain facts are true, without giving enough attention to whether those facts are actually, in fact, supported by the evidence. It is particularly important to ensure that all the evidence supports our facts, rather than just finding some evidence that supports our previously held views or wishes. Conversely, it is important not to ignore evidence that is imperfect but still helps. Scientifically, we can reserve judgement and accept that something is uncertain. In comparison, when faced with an important ethical decision, we should not reserve judgement about what to do – we should make the best decision we can on the evidence available.

A more common weakness is where our reasoning gives only very weak or circumstantial support for our conclusions. Indeed, rarely do we actually provide perfect evidence and logic. The danger is that lots of arguments sound more convincing than they really should be, unless we look into them. So we should carefully analyse what we (and others) say, to see how well it really does defend our views. However, a weakness does not necessarily mean our starting points or conclusions are wrong; it only suggests we cannot be completely confident they are right. Such weaknesses are somewhat like bone fractures in our thinking – the fractures are problematic because they stop the bone supporting the body's weight, or allow too much movement.

Ideas

 Catabolic and anabolic ethics
Some moral reasoning builds up our arguments, reaching new conclusions or strengthening our current reasoning. Other reasoning is more destructive, challenging our views, reducing our confidence and increasing our humility.

 Retrograde thinking
Sometimes we think about things the wrong way round. We may confuse 'therefore' and 'because' (e.g. arguing that because 'animal rights extremists' oppose battery cages, then anyone who opposes battery cages must be an animal rights extremist). We may also combine two views or facts in a way that puts one the wrong way round (e.g. concluding from 'curing disease is always an act of veterinary practice' and 'curing disease is always good' that 'an act of veterinary practice is always good').

Making Decisions

8

8.1 Recognizing biases

Perhaps the most important skill in ethics is identifying possible predispositions that could make us blind to weaknesses, prone to particular errors, or liable to over-represent a particular type of viewpoint. These might be personal biases, for example towards views that justify what is best for us. They might be cultural biases, such as an unthinking acceptance of a social practice (e.g. individualism, capitalism or common animal uses) or an attitude (e.g. preference towards dogs or humans above rats). As with other errors, a view being biased does not necessarily mean it is wrong (indeed, we might think some biases are morally right). However, identifying a morally unjustifiable bias in our own or others' thinking may make us more sceptical about its conclusions.

Biases can be easy or hard to identify. Generally, identifying others' biases is often easier than identifying one's own, so one can try to critique one's own view as if it were someone else's, or ask someone else to pick holes. Common predispositions (e.g. species biases) are also harder to spot, since they do not stand out or get noticed by those who share them – rather like extreme morphologies that become common in a particular breed and then are seen as normal and acceptable. Some may even be identified, but then defended (e.g. that humans are morally more important than any other species), but often in a circular way that may reveal the very same bias.

Ideas

 Fixing the results
Sometimes our challengeable ethical starting points predetermine our answers and preclude challenges (e.g. 'human rights' predetermines that rights are for all humans and only humans). We should instead keep challenging our fundamental assumptions and implications.

8.2   Grafting ethics

Just as we extrapolate clinically from other cases and studies, our conclusions in one case can help us decide what to do in others. We might draw analogies from paradigmatic cases to other cases (e.g. as is common in some religious ethics, legal processes based on precedent, and some forms of business ethics).

However, extrapolation is harder than it sounds. Firstly, we have to be confident about our views on the cases from which we are extrapolating (unless we are provisionally extrapolating to test if those views apply more generally). Secondly, we need to identify what factors should be similar (and how similar) to justify extrapolation, and avoid extrapolations based on irrelevant or biased considerations (e.g. race or species membership). Thirdly, we need to decide which case to copy when our new case could be compared to several different previous cases. For example, if asked to dock a pet cat's tail, do we compare it to the equivalent procedure in dogs or sheep, or to declawing cats, or to genital mutilation in humans?

We might be particularly nervous about extrapolating across very different contexts, as an idea might be incompatible with other ethical views native to that context or might damage them (like a graft might be rejected or carry pathogens). We might also worry that one could use this approach in biased ways to defend almost anything you want, by selecting the right comparisons. We might generate some principles for extrapolation (e.g. we might identify which similarities between the 'donor' and 'recipient' suggest extrapolation will be successful), but then we might as well just make decisions using the principles themselves.

Ideas

 Non-representative sampling
Sometimes we defend a behaviour by citing what is done in other, different cases (e.g. defending tail-docking for dogs by pointing out that we tail-dock sheep). Sometimes, these cases are abnormal and unlikely (e.g. it would be acceptable to eat meat if we were starving, so it is acceptable to eat meat daily). We should think carefully whether the reasoning in one case should apply to another, and why.

 Copying human medicine
Sometimes we use medical ethics ideas (e.g. consent) without fully considering the differences between human medicine (where the client is the patient) and veterinary practice (which involves a three-way relationship). Instead, we should choose our sources cautiously (e.g. focusing on paediatric ethics), evaluate the concepts critically, and adapt them to fit veterinary medicine carefully – as we would do for comparative anatomy, physiology, behaviour, pharmacology or medicine.

8.3   General ethical views

Another skill is to formulate or identify generic ethical viewpoints (which we can then apply to other cases). We can do this provisionally, thinking of our general views as alternative hypotheses to test. If a view seems right in all cases, this suggests there may be universal principles (albeit accepting we always theoretically find new cases that they do not fit). If not, then this might take the general principle to be refuted or needing amendment to fit the new cases.

We might try to identify a 'covering law' for all ethics, by generalizing 'top-down' from one concern, including overall outcomes in terms of experiences (Table 8.1). However, while such an approach might convince some people, we might not be optimistic about finding a theory that everyone (or even every ethicist) believes. Alternatively, we might try to ascertain general rules by considering multiple cases 'bottom-up'. As in veterinary science, ethical theories may be developed from 'data' of many cases, used to formulate generic theoretical ideas, and then applied to particular cases. Ideally, we could combine both processes iteratively – repeatedly amending our views, reapplying them and reflecting further – until we have a complete set of consistent ideas with which we agree in all their applications and can use in practice (Fig. 8.1).

Table 8.1. Ethical covering laws.

Concern	Example theories
Experiences	Hedonic act utilitarianism
Satisfaction of motivations	Direct preference utilitarianism
Impartiality	Justice as fairness
Individual freedom	Liberalism
Basic minimal standards	Animal rights
Common goods	Communitarianism
Rules	Deontology
Character traits	Virtue ethics
Agreements	Contractarianism
Interpersonal relationships	Relational ethics
Care	Ethics of care
Naturalness	Deep green ecological ethics
Individual cases	Particularism; situational ethics
Comparing cases	Casuistry
Logical consistency	Ethical rationalism
Formulating rules with best overall consequences	Indirect utilitarianism

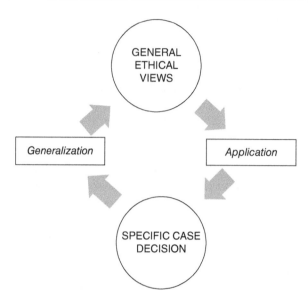

Fig. 8.1. Top-down and bottom-up thinking.

The trouble is that none of our ethical viewpoints do both jobs well. Some are generally uncontroversial but are not very useful in practice; others do help us make decisions but have some counterintuitive conclusions. For example, a general concern for character traits can comfortably capture our views for many of our cases, but is not so helpful in making decisions. Lists of absolute rules can clearly and precisely direct our behaviour, but do not really fit with our intuitions. Perhaps this is unsurprising, otherwise we would all have agreed on an ethical theory long ago. So we might instead think that ethics should be just 'case-by-case', but this would risk ignoring that there are at least some general ways in how we think we should behave.

Ideas

 Over-generalization
Sometimes we might generalize too hastily from limited cases (e.g. concluding, from cases where owner consent is useful, that we always need consent). We might even think that we should not generalize ethical principles and apply them to others at all.

 Confusing 'some', 'only' and 'all'
We sometimes get confused about whether our beliefs apply to 'all', 'only', 'some' or 'none' of a population (for example, owners might recognize that animals expected to experience severe pain should be euthanized but

continued

conclude that their own animal should not be euthanized if it is 'only' in severe pruritus – as if it is only animals in pain who need prompt euthanasia). We should instead ensure we are clear whether our beliefs 'always' (or 'never') or 'sometimes (do not)' apply to a population or case type (Table 8.2).

 Pigeon-holes and hybrids
Sometimes we think a theory sounds plausible and then unthinkingly force all our views to fit that theory, even if it feels counter-intuitive, rather than reflecting whether we might intuitively have spotted a weakness in the theory. In practice, we often have multiple ethical views (e.g. concern for human life and animal suffering). Such approaches may lack logical consistency, but may fit better with our moral intuitions.

Table 8.2. Mixing some/all/only.

Starting point	Mixed-up error
All animals in intractable pain should be euthanized	This animal who is not in pain should not be euthanized [mistakes 'all' for 'only']
Humans have moral rights	Animals have no rights [mistakes 'all' for 'only']
Treatment should not be provided without owner consent	Treatment should be provided with owner consent [mistakes 'only' for 'all']
Novel treatments can be beneficial	Novel therapies should be allowed [mistakes 'some' for 'all']

8.4 Applying general views

One way to reach ethical conclusions is to apply our general views to the specifics of each particular case. We might think of each ethical view as having a particular scope (Table 8.3). Each view has inclusion criteria to which it applies, or which 'activate' it (rather like how receptors engage with particular proteins). Sometimes these form a general ethical statement (e.g. we should help suffering dogs) and might engage particular factual statements (e.g. this dog is suffering) to form an applied conclusion (e.g. we should help him). Each view also has exclusion criteria where it does not apply. For example, all moral responsibilities only apply in cases where we can actually fulfil them – we have no duty to do the impossible.

Sometimes the scope of a viewpoint is fairly obvious (e.g. motivations against causing pain apply to behaviours that cause pain). However, many of our ethical views prove to be quite difficult to apply 'at their margins'. For example, it can be difficult to precisely define ideas like euthanasia, murder, harm, help, lie, defraud and steal. Even biological categories are surprisingly

Table 8.3. Inclusion/exclusion criteria.

View/motivation	May apply for	Does not apply for
We should allow/ perpetuate enjoyable lives	Animals with good prognoses	Futile treatments Treatments that perpetuate suffering
The responsibility to fund treatment is the owner's (not ours)	Owned animals	Wild animals or strays needing emergency treatment

difficult to tie down (indeed, species are not as hard and fast categories as many people think they are). For example, a principle that applies to humans seems quite straightforward, but raises questions about embryos, fetuses, chimeras, hybrids, other anthropoid hominids and great apes.

One approach is to consider whether a fact is sufficient, necessary, or both or neither, to lead to a particular conclusion. We might think an ethical view applies whenever a particular fact seems true (e.g. we should legally protect any animal who can suffer pain); or does not apply when it appears false (e.g. animals who cannot suffer pain should not be legally protected) or both (e.g. animals deserve protection if, and only if, they can feel pain). At other times, we might think multiple factors are neither necessary nor sufficient, and instead should be considered together and cumulatively (e.g. whether we should kill an animal depends on multiple factors, none of which is, by itself, determinative). Knowing if necessary, sufficient, or both or neither applies is therefore very important – and people often get confused and thereby mix 'some' and 'all' or 'none' (e.g. applying a concern about all humans as if it applied only to humans).

We might also think of the scope of one moral view as being limited or 'eclipsed' by other concerns (e.g. our moral motivation not to restrict animals may be limited by another motivation to avoid other animals getting infections; similarly, we might think we should follow clients' wishes, or fulfil promises, but not where their content is immoral). We might even think our moral concerns are limited by those of other people (e.g. we might think that we should treat owned animals to avoid pain, but not where owners refuse to fund it). Such limits might themselves have scopes. For example, we might not exclude cases where there is no owner (e.g. emergencies or unowned animals) or for basic first aid (e.g. euthanasia).

Ideas

 Constraints
We might think one moral responsibility should be constrained by particular practical or moral factors (e.g. we should only help our patients where this does not harm others).

continued

Ethics-by-numbers
Sometimes we espouse or apply an ethical viewpoint indiscriminately, even to cases that should be exceptions (e.g. withholding information even when doing so would prevent animal abuse). We should instead be clear when any moral principle should and should not be followed.

8.5 Conflict resolution

Our ethical deliberations are likely to identify multiple concerns that apply to a particular case, especially if we are skilled at identifying a wide range of factors and ethical views (it is easy to identify one principle and apply it to every situation – and then have unsurprisingly strong views on what should be done). The greater skill is being able to work our way through multiple clashing principles. There are several different approaches to resolving such conflicts. One (which we have covered) is to see if the scope of one concern is limited so that they do not apply.

Another approach is to identify an underlying ethical concern that is fundamental to both. For example, we might worry about vaccination risks but also about infectious diseases. In this case, the underlying ethical concern is to minimize risks of illness, so we can then select the approach that would best reduce the overall risk profile. However, we might find that some concerns cannot be reduced to a common denominator.

Another approach is to decide which moral concerns are more important (in each particular case) and treat them as 'overriding' our other motivations. In some cases, one motivation is a priority above all the others (e.g. we might think preventing suffering is more important than fulfilling clients' wishes). In other cases, we might 'weigh up' all the factors to see which is more important in a given case. For example, if an owner has two old, bonded animals and one is severely ill, we might decide the benefits of euthanasia to that animal are outweighed by the harms to the other of isolation (or vice versa).

Ideas

 Feeling inconsistent
Sometimes our beliefs feel inconsistent with each other or with our behaviour. If we spot the inconsistency, we might modify our views or our behaviour in future.

 Duck out of hot water
We sometimes answer moral questions in ways that avoid the main issue (e.g. we might decide to persuade a client or colleague to change their mind – but ignore the trickier ethical question of what to do if they still do not change). We should instead tackle all aspects of a moral dilemma.

8.6   Ethical palliation

Unfortunately, we might not always feel able to resolve conflicts and confidently identify the right option. Sometimes, our moral concerns suggest no
option is right or best. We might feel it is wrong to provide analgesia that a
client has forbidden, but also wrong to leave an animal suffering. Some of
these situations seem, sadly, just due to bad luck. Many of them are, frankly,
the result of other people acting unethically. Often, such dilemmas have been
caused by our own previous wrong behaviour, where we have already missed
the chance to do what is right. For example, if we have undertaken to provide a treatment option that we later reckon will harm our patient, we might
then think it wrong to provide it and wrong to renege on that promise.

We can get better at focusing on the decision we have to make, and not
being distracted into lamenting the situations and how other people have
caused them, or thinking wistfully about how the case might have been different. We can think about how we might improve the situation, even if we cannot
make it perfect. Indeed, veterinary work is often about making the best of
bad situations. We can also recognize why these are bad situations and try to
prevent them in future (e.g. by helping owners to treat their animals differently).

Ideas

 Moral distress
We might feel a particular type of anxiety when faced with moral
dilemmas (in addition to the confusion and stress caused by any difficult situation).

 Making the best of it
Lessening suffering in horrid cases can feel unrewarding (compared
with achieving something amazing), but it is just as (and often more)
important.

8.7   Making the decisions

Pulling together all the skills we have considered so far, we might try to ensure
our decisions are sufficiently robust across several criteria (Table 8.4). In one
sense, this is the key skill of ethics.

This does not mean we should use burdensome decision-making processes every time we are faced with a choice. We should engage in more
robust decision-making for difficult, contested, novel or important cases.
We should also be more careful when we might need to overcome – as best
we can – our potential predispositions and biases due to habits, pressures,
self-interest or ignorance. This means explicitly considering the factors,

Table 8.4. Criteria for robust decision-making.

Criterion	Question
Internally consistent	Is our decision logically supported?
Externally consistent	Does our decision fit with our other related decisions?
Comprehensive	Does our decision take all relevant facts and values into account?
Accurate	Is our decision based on accurate facts about the real world, and correct interpretations of underlying principles?
Compatible	Does our decision fit with our fundamental, unchangeable intuitions?
Determinative	Does our process actually help make a concrete decision?

reasons and logic in our decision – to be able to explain our rationale to others and also to ourselves. It also means being extra careful when making decisions that could have a major impact on others or where we have historically always acted in a particular way without really thinking about it, which can make it hard for us to notice errors or challenge ourselves.

We then make our decision from the available options. Biologically, we might think of ourselves as having various ethical motivations towards or against each option. We might be morally motivated to help our patients, to avoid harming our clients, etc. Our ethical motivations may have different strengths. Some can outweigh others; for example, the motivation to avoid causing surgical pain may be outweighed by a motivation to prevent greater pathological pain from a fracture. Occasionally, the strengths of different motivations may reflect factual matters (e.g. varying severity of suffering). Sometimes, we can ask humans or observe animals' preferences to see what outcomes (or trade-offs) are more important for them. However, there are no scientific units of ethical motivation (or suffering or health), so we ultimately need to make a decision based on our professional ethical judgement.

One biological way to consider our ethics is to think of our different ethical reasons as 'activatory' or 'inhibitory', as regards our motivations. Activatory views may motivate particular behaviour (e.g. avoiding pain might 'activate' a motivation towards medicating). In comparison, inhibitory views weaken or neutralize other motivations. For example, uncertainty over a surgery's efficacy or worries about side-effects might make us less motivated to provide the surgery. Inhibitory views can also inhibit inhibitory reasons (just like inhibitory neural pathways can inhibit inhibitory pathways). For example, we might consider that we should be less concerned to avoid side-effects in patients who will be quickly euthanized if any do occur. On their own, inhibitory motivations can only steer us towards inactivity (i.e. doing nothing) – we need some positive activatory motivations towards particular behaviour if we are to decide to do anything.

Ideas

 No excuses

Sometimes we fail to recognize our responsibility and instead consider ourselves overly constrained or predetermined by others, or by factors we could actually control or ignore. We should take full responsibility for making and fulfilling ethical decisions, without blaming circumstances or others.

 Two negatives make a positive

Sometimes we base a decision on the absence of arguments against it. For example, an animal not being in pain is not a reason for keeping an animal alive (it is merely the absence of one possible reason for euthanasia). Similarly, an animal's inability to consent is not a reason to permit research or treatment without consent. We should still try to make decisions based on positive activatory motivations.

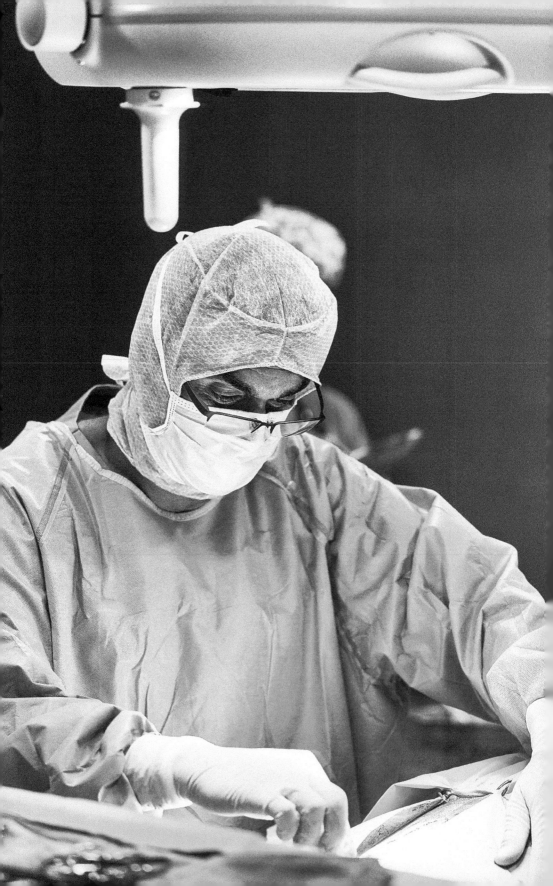

Getting It Done and Getting Better

9

9.1 Implementing Our Decisions

Once we have made our decision, we should implement it. However, implementing decisions can still be difficult and emotional, and we may even sometimes fail to do it (e.g. due to pressure, emotions or forgetfulness). Implementing our decisions may also lead to unintended outcomes (e.g. surgical complications, or unhappy clients and colleagues), which constitute new ethical decisions (e.g. to provide analgesia or euthanasia, or to apologize), and also make us question ourselves.

We might improve our confidence and resilience by knowing we have thought logically, comprehensively and selflessly. We might also try to improve our follow-through by developing 'moral habits', or 'nudging' ourselves by making public commitments and setting time-frames. We should also ensure our ethical decisions are achievable – without giving up our ambition to do what is right. Finally, recognizing our limitations should help us to be humble about our moral authority and non-judgemental about others.

*

In implementing our decisions, we should also ensure that our secondary decisions are followed through, i.e. our conditions are met. For example, if we evaluated surgery as the best option on the assumption of effective analgesia, we should provide it. This can often require us to ensure other people do their part in achieving the desired outcomes. This involves ensuring they are able and willing to do so (Table 9.1).

Ideas

🧠 *Not doing your own dirty work*
Sometimes we make decisions that others have to implement, in ways that mean we avoid understanding or engaging with the issues (e.g. as consumers who never see farming or slaughter).

---

continued

Moral incontinence

Sometimes we fail to do what we have decided when the moment comes to do so. This might be for new reasons or emotional factors that we had not previously considered and which made us review our previous decision. But sometimes it feels like we just fail to do what we still know is right. Logically, for completely rational people, such weakness is impossible.

Table 9.1. Abilities, opportunities and motivations.

Factor	Examples
Knowledge	Case details
Skills	Pain assessment
Opportunity	Sufficient time
Resources	Equipment
Social support structure	Practice team
Motivation	Patient compassion
Lack of competing motivations	Lack of personal interest bias
Memory	Prompts

9.2   Case reviews

Considering other cases might make us reconsider our conclusions in the original case from which we extrapolated. For example, applying our conclusions to other cases could imply we should take options that we might think immoral or absurd (e.g. our belief that animals' cognitive limitations make it acceptable to farm them for meat might also suggest that it would be acceptable to use human children with learning difficulties, prompting us to question that original belief). We might use such comparisons to look for irrational prejudices that we should overcome, or for other legitimate moral concerns that we should take into account. One benefit of experience is having seen a wider range of cases to compare and contrast.

We should ensure we reflect on our previous decisions, to see what we can learn. This can be difficult and unpleasant – particularly if we find we made an error – but it is vital. The only thing worse than making a mistake is to keep repeating mistakes. We should avoid an excessive bias towards defending (and repeating) our previous behaviour, even when we have already sunk time, money or emotion in them (e.g. whether we have already provided extensive treatment for a patient is not a reason to provide more). We should do things because we can defend them, rather than defend them because we did them. We can get better at critically reviewing our previous behaviour, while forgiving ourselves for previous errors or accepting we made the right choice at the time. However, we should also avoid sentimental self-recrimination for past behaviour.

Looking forwards, we should try to predict subsequent moral dilemmas that our behaviour could create, to avoid putting ourselves in them. For example, by offering options to a client that we think are unethical because we feel we should give them the 'full range' of available options, we might then put ourselves in a difficult position to refuse an option they have selected (and potentially offending them by implying we feel they have made an immoral choice). Or we might place ourselves in a moral dilemma by offering treatments to an owner that we think would harm our patient, so might decide that we should only offer treatments we consider morally justifiable (which might be a range, from which owners could choose).

Ideas

 Doggedness
Sometimes we find justifications for what we have already decided or done, and even increase our commitment. We should instead critically reflect on our previous behaviour, and possibly change it in future.

 Pride
We sometimes think we are already ethical and do not need to change. We should instead always be open to change.

9.3 Ethical explanations

Another skill is being able to explain one's ethical views in particular cases. This involves identifying what facts, values and reasoning were relevant to one's decision, how they were applied, and why other factors were discounted or outweighed.

We should be able to justify our behaviour (i.e. explain why it is right). However, we should avoid trying to justify whatever one has thought or done uncritically, starting with the assumption that it was right and finding ways to defend it post hoc. Indeed, skilled ethicists can find potential reasons to mount a defence of almost anything. Instead, we should look for genuine reasons, critically assess them and check that they are 'justified justifications' (and that their justifications are justified, etc.). In our justifications we should be open to finding that we are not justified.

We should also be able to justify our general ethical views, in terms of why we think they are important. This is partly about explaining why other ethical concerns are not important, to identify what are our core concerns. This might draw on some wider beliefs about what justifies an ethical view (e.g. that it should be applicable to all relevant cases; that it should be logically consistent; that it should fit with our intuitions after due reflection). It is also partly about expounding those concerns so that others can understand them. This might draw on various methods such as restating or 'unpacking' the views, giving examples and clarifications, and comparing them with superficially similar ideas, or describing the basis, reasoning (i.e. showing our working) and implications of our views. This combination of refinement and explanation (like descaling and polishing) should leave us with clear, defensible views.

We might find we can explain or justify our views and behaviour to ourselves fairly easily. We might find it harder to explain or defend our views to others, especially if they have fundamentally different ethical or factual assumptions. Sometimes, we might provide explanations that the other person considers irrelevant (e.g. if they consider ethics is about one's own phenotype, then they may not consider our concern for outcomes to justify what they see as wrongdoing). Sometimes, we might think we are agreeing, but actually we are understanding concepts in very different ways.

Ideas

 Assuming authority
Sometimes we might simply assume that we, as vets, have some moral authority on animal issues due to our factual knowledge. Instead, we should ensure that we listen to others' views, and base our own moral authority on deep ethical reflection and brave, selfless views.

 Ambiguous language
Sometimes we confuse the multiple meanings of words. For example, barbiturate overdose is humane in the sense of 'painless' but not necessarily humane in the sense of 'kind'. Other ambiguous words include 'good', 'right', 'responsibility', 'fair', 'professional', 'animal rights', 'animal', 'irrational', 'person', 'humanity', 'emotional', and words that have both technical and everyday meanings (e.g. 'stress'). We should instead use words precisely and unambiguously, and ensure everyone understands exactly what our words mean.

 Talking past one another
We sometimes focus on facts that are relevant to our moral views, but not to our audiences. For example, owners concerned about killing (a behaviour) may not be persuaded by poor prognoses (an outcome). We should instead engage with others' fundamental ethical assumptions.

9.4   Self-improvement

One final skill is being able to get morally better, for others or for our own flourishing. We might look for an epiphany or life-changing event that radically changes our outlook on morality. However, it can often be hard to implement change in ourselves, since we have habits and outlooks formed by years of operant conditioning. Alternatively, we might try to change ourselves bit by bit, developing new habits and character traits by practice – bolstered by self-monitoring, reflection and possibly feedback from others.

We can improve our skills by experience, but this 'secondary intention' can be slow and painful. Alternatively, we can develop by debriding our views, and reconstructing them together, so our resultant viewpoint 'tissue' is stronger and has greater integrity. This means we should be open to changing our mind and our behaviour. We should accept all our ethical views as being open to challenge and improvement. We should be open to continuous development, and avoid any conscious or subconscious resistance to change. What we have done (or not done) is now done. We cannot go back. What we can change is what we do in future.

This means we should practise being open to new ideas, or even to views that we previously found unappealing. We may react to ideas excessively or with hypersensitivity, guarding ourselves from any contact (sometimes aggressively) and avoiding or rejecting related ideas. We can have allergic reactions to some ideas (e.g. 'animal rights'), where we fight them but might actually find them commensal or symbiotic with our current views. Sometimes we even attack our own views (in a sort of autoimmune response). Instead, we should try to reduce our immediate responses, so as to evaluate whether the ideas can – more like food – be taken in, digested and assimilated, or excreted once we have gained all the ethical nutrition we can. (To stretch the analogy, ethical reflection acts like villi – improving the amount we can take in).

Ideas

 Getting perfect
We might think of our moral improvement as part of (rather than opposed to) our self-interest and as part of our flourishing.

 The best we've got for now
We should accept, expect and hope that our current ethical views – like our scientific beliefs – will improve over time.

Part C: Applying Ethics to Veterinary Work

There is little value in having ethical viewpoints if we do not, or cannot, apply them usefully to our real-life decisions. This section considers how we might apply our reflections to key areas of veterinary work: clinical practice and practice management; veterinary research and education; veterinary policies; and engaging in wider ethical debates. We sometimes distinguish between ethical theory and practice. However, ethical views that we think do not fit reality are ones that we reject as wrong (just as we would reject a scientific theory that did not explain or predict evidence). We should revisit our theories until they fit what we think is actually right in practice.

In this section, rather than considering our previous cases, we will try to predict or imagine future ones. Even better, as we meet cases in our work, we can start to apply the ideas (and avoid the errors) that we have developed so far. Trying to predict future challenges can act like a vaccine – helping us to develop our responses to challenges safely (and without feeling ill), so that we are more ready to respond in our behaviour, and also feel somewhat inoculated against the emotional effects of such cases (although this does not mean we are completely immune). We can also try to anticipate controversial or difficult cases and imagine ones that are outside our normal experiences, as such features can act like an adjuvant in stimulating our engagement.

This imaginative approach can also help us avoid being too defensive, focused on the law, or distracted by irrelevant facts. Indeed, we might want to consider our cases more objectively, conceptually and open-mindedly by imagining they involve completely new clients in another country, where we are unfamiliar with the culture and law. At the same time, if we are working as vets while reading this book, we can consider the cases we actually see. We can also use our real and imaginary cases as 'experiments' to explore what affects our moral views, and to test hypotheses about what is right. We can use simple, contrived scenarios (e.g. whom to save in an emergency) to isolate 'variables'; and more complex and realistic cases can help analyse practical decision-making.

Clinical Veterinary Practice 10

Clinical ethics is a clinical skill. It is a key part of clinical practice, in specific cases and in our practice management. Each aspect of practice has its own challenges and questions (which apply whether one is in equine, farm, laboratory, official, small animal, wildlife, zoo, or any other work).

10.1 Diagnosing and prognosing

In practice, we investigate to find out information (e.g. clinical examinations and history-taking that gain information about an animal's present and past states and contexts). Such methods can sometimes cause pain (e.g. sampling) or distress (e.g. hospitalization), delay treatment, use up resources that could be spent on treatment (e.g. insurance limits) and create anxiety for owners, and sometimes we obtain information upon which we cannot usefully act.

Let us anticipate a clinical case where we could do various tests. How should we conduct our clinical investigations and assess options?

- What makes investigations valuable? How do we weigh up tests' benefits and risks (often where they are poorly quantifiable)? How do we use information about a patient's history and current state to decide what to do next?
- How should we conduct investigations to help us consider multiple possible outcomes? Who should perform these assessments?

We might think diagnostics are valuable because (and if) they are informative. However, our diagnostics concern a patient's history or status, while our ethical and clinical decisions concern the future. Alternatively, we might think diagnostics are valuable only if they improve our decisions and clinical outcomes. With this approach, we might think our investigations should focus not on the past (aetiologies) or present (disease state) but on predicting future potential outcomes for different treatment options (although information about the past and present is obviously useful for making such predictions and for evaluating treatments going forwards). This character-izes diagnoses less as descriptions of states and more as prescriptive nodes in our decision-making. Either way, we should weigh up such risks of any diagnostic test (e.g. pain, delay) with the expected chance of improving our decision-making: when we select a diagnostic test, we are making a 'bet' on what we think will benefit our patient.

*

We might prognose future outcomes using our intuition, since we have no scientific 'crystal ball' (and no scientific units for intensity). However, this risks undisclosed biases and limits meaningful discussion. Alternatively, we might think we structure our assessments, for example by evaluating the intensity, duration and probability of different possible periods of enjoyment or suffering, scoring intensity in terms of trade-offs against durations (e.g. equating one hour of severe postoperative pain with one day of mild hunger; or two months with mild arthritis with one month of normal healthy life). However, this requires very complex and uncertain assessments, unless we simplify our assessment (e.g. to one year and ignoring negligible-risk out-comes). In between, we might decide that the level of detail and structure should depend on the perceived degree of risk (e.g. being more detailed and structured for novel procedures or complicated cases).

We might think outcomes should be assessed by us veterinary profession-als, as we have better general knowledge of patients' biology and the poten-tial effects of treatments. However, we usually do not know our patients as well as their owners do. Alternatively, we might think we should let clients evaluate options for their animals, whom they know well as individuals. However, clients might have limited understanding of science or medicine, and might have personal or emotional biases, erroneous preconceptions or cognitive incapacities. So we might think we should assess options together, sharing our insights. However, this does not help us decide whose evaluation to use when we and owners disagree.

Ideas

 Imposter syndrome
Sometimes we feel as though we are unqualified to make decisions or give opinions, but this should not mean we just leave those decisions to others who are even less well qualified (e.g. many owners).

 Mistaking 'can' for 'should'
Sometimes we (or owners) might struggle to decide not to provide a treatment that is available (in terms of technology, skill and finances). Instead, we should recognize that, even when a treatment is possible, we still need to decide whether it is the right one.

continued

Overtreatment

Sometimes we want to fix patients, or find out what is wrong, so much that we are overly predisposed towards interventions or investigations whose chances of benefits are outweighed by suffering or risks. We should only provide treatments that are better than no treatment, more conservative treatment, or euthanasia.

 What killed the cat

Even 'scientific' diagnostics can involve feelings of curiosity, relief, satisfaction or pride about reaching diagnoses. We should avoid performing diagnostics purely in order to make ourselves or owners feel better; we should do so only to improve decision-making.

Applications

- What can our patients' past and present tell us about the possible future?
- What are the possible outcomes for each treatment option (including no treatment)?
- Would finding out more information risk harming the patient (e.g. delays or iatrogenic suffering)?
- Are we unduly biased towards getting a diagnosis, more than that diagnosis is actually worth?

10.2   Selecting

We provide veterinary treatments to alleviate patients' suffering, extend their lives, and help owners and others. However, sometimes the best treatment for our patient might have unintended effects for the patient or others (Table 10.1).

Let us progress our imaginary case so that we have various treatment options (these may be surgical, medical, nutritional, behavioural or environmental), including euthanasia (which is arguably an option in all cases), some of which are better for our patient but have other effects on other animals or people.

- How should we resolve conflicts and limit the help we provide?
- How should we decide treatments if we only consider our patients? When is it right to extend or shorten a patient's life? Does that depend on what else we could do?
- How do, and should, owners affect our decision-making? How can we tailor our communication to their ethical views?
- Should we ever harm, or cause suffering to, our patients when that would benefit other animals or people? Conversely, should we always help our patient even when it could harm our clients, or other animals or people?

Table 10.1. Side-effects.

	Direct effects	Indirect effects
Patient (and the owner's other animals)	Medicine reactions Cosmetic or performance-enhancing treatments	Extending lives-worse-than-death Preventive mutilations or antimicrobials that mean owners can continue using methods we think harmful
Other people or animals	Blood donation, kidney transplantation Spreading infectious or inherited conditions Causing owners worry, grief or guilt	Reducing herd immunity Spread of antimicrobial resistance Using up resources that otherwise would have been used on needier animals (or people) Production of products that involve harmful animal research, testing or production methods

Table 10.2. Beneficial and harmful treatments.

	Pain/suffering/deprivation	Enjoyment/pleasure/satisfaction
Prevent/reduce	Benefit	Harm
Cause/encourage	Harm	Benefit

For our patients, we might think that we should never cause suffering (or death), but this would prevent even beneficial treatments that involve risks (which is probably all treatments). Alternatively, we might think we should ensure any potential risks are outweighed by the benefits so that each patient gets the best 'treatment bet', compared with other options (including no treatment), considering both positive and negative outcomes (Table 10.2). Either way, we might think we should minimize any 'iatrogenic harms' that we cause through our investigations and interventions (e.g. through analgesia and in-patient care, or briefly delaying euthanasia to help the owner cope), except where this would itself risk greater harms.

We might decide we should keep all patients alive as long as possible. However, some lives are 'lives-worse-than-death' for the animal. So, alternatively, we might decide we should only keep alive (and not kill) patients whose chances of enjoyment outweigh the risks of suffering, so that a longer life is the best 'bet' for them. In comparison, we should kill (and not keep alive) patients who would be better off dead than continuing their suffering from their condition, management or clinical treatment. This suggests we should actively cause death rather than letting animals die in ways that involve more suffering.

*

Which option is the best bet obviously depends on what other options are available and morally acceptable. Sometimes other options are unavoidably ruled out (e.g. euthanasia may be the only possible way to avoid lives-worse-than-death in severe incurable illnesses). Sometimes, options are effectively ruled out by other people (e.g. clients refusing to provide outpatient treatment). We might think we should refuse to provide treatments that are indicted only because of an owner's moral failures, since the responsibility for that suffering is the owner's (and we do not want to endorse or facilitate their failures), but this risks allowing suffering. Alternatively, we might pragmatically decide to provide whatever treatment is the best bet, however lamentable the constraints (and not act like we are in some imaginary better world). Oddly, this can mean what is contextually right for us is wrong for owners (e.g. if an owner refuses to change husbandry methods, prescribing a mutilation might be right). This consideration may help us avoid feeling guilty for owners' failings.

*

We might think we should provide whatever treatments benefit our clients (regardless of outcomes for our patients), if we think owners should have control, or human desires are more important than animals' needs. Conversely, we might think we should do whatever helps our patients most (regardless of outcomes for anyone else). However, both approaches obviously might involve harming others (potentially severely) in some cases, and could reduce our reputation as a caring profession (thereby leading to worse outcomes overall). Alternatively, we might think we should provide whatever treatment leads to best overall outcomes across everyone affected, but this seems an impossible calculation and might sometimes suggest we provide treatments that harm our patient and client to benefit others, which we might think is a betrayal of that relationship.

As another alternative, we might think we should avoid causing harms. We might think we should avoid harming our patients overall (i.e. refuse to fulfil requests when the benefits are outweighed by the risks). We might further think we should avoid treatments that harm anyone. However, this could be very restrictive, since almost every treatment poses some risks to someone (e.g. in driving, or using electricity). We could limit our concern to avoiding harms that are deliberate, likely, disproportionate and/or primarily our fault (Table 10.3). However, this may make us feel like we are 'washing our hands' of harms we know are associated with our actions.

Table 10.3. Harms for which we might feel responsible.

Harm to avoid	What this might still allow us to cause
Deliberate	Accidental harm (e.g. treatment side-effects)
Realistically likely	Negligible risks (e.g. remote possibility side-effects)
Greater than the benefits to our patients	Where benefit to patients outweighs harm to others (e.g. blood transfusions)
Harms for which we are primarily responsible	Harms that are 'others' fault' (e.g. drugs and meat, if farmers and pharma companies have not cared for their animals well)

We could also ensure our clients or patients do not seriously harm any-one else. As examples, we might decide we should not release animals who we would expect to pass on serious infections or harmful genes (and we might then decide that euthanasia is a better option than permanent con-finement), or help our clients to continue poor or polluting management, neglect or abuse (e.g. by reporting them). However, this approach would suggest not helping any predators (including hunting pet cats), unless we modify our rule not to help patients harm anyone else in unnatural ways.

Ideas

 Fool's gold
It can feel disappointing and frustrating not to provide the very best treatment to every patient, especially when this is due to others (e.g. owners not being willing to pay). However, this attitude might miss opportunities to help patients (albeit to lesser degrees) and we should still feel good about helping animals as much as we can, and thereby making them better off than they would have been without us.

 Euthanasia
Euthanasia refers to causing or allowing an animal's death to prevent an expected life-worse-than-death (and minimizing the suffering in the death process).

 Reversibility reservations
We sometimes think of some treatments as irreversible (e.g. euthana-sia) and others as reversible. But all options are irreversible – we can-not undo any harms we have allowed or caused (although we might sometimes be able to compensate others for them).

 Assuming win–wins
We may sometimes naïvely reason as if the world and other people are perfect (e.g. as if owners always want what is best for their animal). Instead, we should avoid over-optimistically limiting our thinking to such cases, and still focus on potential dilemmas.

 Prioritizing our patients
One approach to veterinary ethics is to focus our decisions on doing what is best for our patients' welfare within particular constraints (e.g. following the law). This includes both clinical decisions and practice management (e.g. policies and business models).

Applications

- Would this treatment harm, or risk harming, our patient, our client or others?
- Can we reduce the risks or suffering involved (without causing worse harms)?
- Can we legitimately consider some risks as negligible or outweighed?
- Would our patient otherwise have a life-worse-than-death? What suffering might occur in the process of death? Can this be reduced? Does it mean no treatment is actually better for our patient?
- Which options are better than no treatment, more conservative treatment or euthanasia? What option would this patient want as an 'idealized preference' or which seems the 'best bet'?
- If there are other theoretical options, are they ruled out so that we must choose whatever is the best bet in reality?
- Might our thinking be biased due to our own interests or pressure from a client? Would removing that bias alter our decision?

10.3 Consulting

Practically, clients facilitate their animals' treatment (e.g. providing transportation, handling, funding and administering medicines); they provide information useful for assessing outcomes (e.g. providing first-hand observations and empathetic insights regarding their animals) and their behaviour can affect clinical outcomes (e.g. whether they follow treatment plans or present animals for further consultations). In some cases, and in some countries, owners' wishes may be legally important, or may be legally overruled by state rules. Sometimes owners might not support the treatment that we consider best, unless we influence their views (e.g. by providing information about expected suffering, or sharing our ethical judgements of 'what we would do').

Let us imagine that, in our case, our client is worried about whatever treatment we thought was best for our patient, and instead is considering another treatment that we consider would cause more suffering. How should we engage with their thinking?

- (When and how) should we influence clients' decision-making?
- If we cannot influence them, who should make the final call on what treatment that animal receives? When an owner wants a treatment, should we always provide it? When an owner does not want a treatment, should we never provide it?

We might think clients usually know best what is good for them (or at least better than we do). So insofar as we want to help, or avoid harming them, we should simply ask them. In comparison, they do not always know or want what is best for our patients, due to personal limitations or biases.

In such cases, we might decide we should help them choose what is best for their animals, although such influence will not persuade every owner. We might also tailor our communication to help owners make robust decisions for their animals, by correcting any misinformation, logical weaknesses or biases (e.g. for euthanasia; Table 10.4).

Alternatively, we might think we should avoid influencing clients' decisions, since this reduces their freedom or risks them selecting options which provide less committed support (or feel guilty afterwards if risks materialize). However, it seems impossible never to influence owners at all (unless we never even provide information) and sometimes to be contradictory (when they request advice). So we might limit our concern to certain methods

Table 10.4. Owners' potential moral concerns over euthanasia.

Concern	Examples	Tailored ethical communication
Personal interests	To avoid grief or inconvenience	Death will happen anyway – euthanasia merely changes the timing, process and experiences involved
Unwillingness to resource other treatment	Inability to pay for/ comply with other treatment options	Tailored treatment plans to budget (or offer of rehoming where in animal's interest)
Desire to 'do all they can'	Belief that spending more shows greater love	'Amount' of treatment does not mean a better owner or outcome
Desire to avoid making a decision	Emotional paralysis	We can help and support their decision
	Belief that their relationship should preclude making fatal decisions	This decision is part of their responsibility
Opposition to active euthanasia	Preference to let animals die 'naturally'	Euthanasia is an act of kindness (and may be legally required)
Nervousness of being wrong	Concern that euthanasia is irreversible	All decisions are irreversible (e.g. pain cannot be un-suffered)
Desire to preserve life, regardless of its quality	Unconditional value on life	Life is not valuable to the animal if full of suffering
Desire to retrospectively justify previous decisions	Not wanting to imply previous expense or effort is 'wasted'	These are sunk costs; it is worse to 'throw good money after bad'
	Not wanting previous decision against euthanasia to seem wrong	What was right then is wrong now

(e.g. physical coercion or threats of violence) or cases (e.g. only for owners who would otherwise cause severe suffering).

*

Where we cannot, or should not, influence clients' views, we might take several approaches (Fig. 10.1). We might think we should do whatever owners want, as they are paying and own the animals. However, this would mean supporting or allowing their animals to suffer. We might aim to make decisions jointly, through open and respectful engagement, to help patients. However, this might well not work in all cases, especially if clients are not primarily motivated by what is best for their animals. We might think that ultimately we should make our own decisions on what to do for our patients, taking into account owners' understanding of their individual animals alongside our generic and scientific knowledge, and not provide any treatment that we think unethical (e.g. performance-enhancing overtreatment or routine prophylactic on-farm antibiotic use). Or we might think we should both retain some control.

Similarly, when owners do not want a treatment, we might think we should not provide it, because this would involve harming them, or because we think owner vetoes generally safeguard animals against harms from vets. Alternatively, we might think we may sometimes provide treatment without permission in at least some cases when doing so would prevent suffering, in particular when owners' preferences cannot be determined (e.g. owners cannot be identified or contacted in an emergency, or when their animal is brought to us as a stray), when clients seem incompetent to make decisions (e.g. when animals are owned by children, or by adults with severe learning difficulties), or when we have another moral responsibility to prevent suffering (and a legal justification, as discussed below).

Combining these approaches, we might think we should provide owners with what we consider to be reasonable options, allowing them to select what they think is the optimal one.

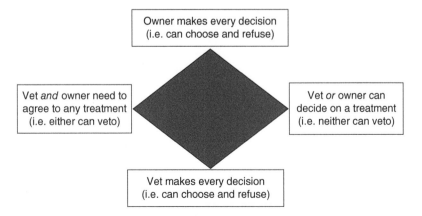

Fig. 10.1. Options on who makes decisions.

Ideas

 Permission

For an owner to give us meaningful permission, we might think they need to understand the treatment options and their implications, be able to make a decision, and be free to make it. Having permission is not a reason for treatment, but its absence might be a reason against treatment.

 Mixing requests, instructions, permissions and authorizations

Permission and authorization may be necessary conditions for treatment (insofar as their absence may be an exclusion criterion or contraindication), but they are not sufficiently positive reasons to provide treatment (i.e. we also need a positive reason and the absence of other reasons against). So, for example, owner permission is not a reason for treatment, even if its absence might be a reason against treatment.

 Under pressure

Dealing with clients and owners can be one of the most stressful aspects of clinical work, as clients may have different ethical viewpoints or motivations to ours, and even try to 'guilt-trip' us. This can make us doubt our own conclusions and feel embarrassed or anxious. We should take clients' and owners' ethical views and outcomes into account in our decisions where it helps us make the right decisions in the circumstances, but we should never be pushed into what we think is wrong.

 Breaking news

We should also consider owners' emotions, trying to minimize any suffering that we cause them in the course of our work. We should never be less than honest – we should always 'tell it like it is', but the way we present the message should be sympathetic (e.g. delivering the news with sensitivity, and signposting bereavement support services).

 Momento mori

Death can cause loss, grief, guilt, anger, denial, fear and a sense of failure for owners – and for us too. We should also try to recognize its value as a final act of kindness (where appropriate). Sometimes, recognizing this love can help us choose the 'best bet' (e.g. briefly delaying euthanasia so that owners can 'say goodbye' might also ensure they make that decision, as well as benefitting them).

Applications

- What does our client intuitively think would be the best treatment for their animals?
- Is the client sufficiently knowledgeable, empathetic, selfless and compassionate to be a competent judge of what is best for their animal?

Should we be sceptical about their information or assessment here? Can we improve their knowledge, understanding or decision-making ability?

- If we cannot agree, do we have another moral (and legal) duty to prevent suffering? If so, does this override or provide an exception to the rule regarding owner permission? Do we have a way to do this morally and legally?
- For strays, would waiting for an owner mean we can better evaluate treatment options? If so, are the benefits of waiting greater for the patient than the suffering due to the delay?

10.4 Charging

Veterinary work needs resourcing. Even advice comes at a time cost. Most veterinary practices are private businesses with equity holders; some are charities funded by others' donations. At the same time, some clients may be unable or unwilling to pay for treatment, other clients are themselves charities, and some animals are unowned or unclaimed (e.g. wildlife and strays).

Let us imagine that our client asks us to provide the option we think best at a discounted price (or free). What should we charge?

- Who should pay for veterinary treatment? (How) should we influence clients' financial decisions on what treatments to pay for? (When) should we collect fees?
- When owners will pay, how should we set our fees? How does this apply to insured patients?
- How should we resource treatment, and at what levels, for unowned animals? Should we provide this free or only if someone else pays?

Some people might think we, as vets, should provide free treatment at the practice's expense. However, we might think this would be unfair when it means that paying clients are effectively subsidizing other owners, and worry that it risks encouraging owners to get animals whose treatment they cannot afford. Others might think veterinary care should be funded by the government (as human healthcare is in some countries), at least for public goods such as controlling zoonotic infections. However, this seems unfair for goods that do not benefit everyone (e.g. if vegans' tax payments contribute to costs due to farming). Alternatively, we might think owners (and indirectly consumers) should fund veterinary care. However, we might recognize that not all owners can or will fund veterinary care (and some animals are unowned, to whom this logic is inapplicable).

We might think clients should be allowed to decide whether they spend money on veterinary care (whatever we decide about providing treatments), as it is their insentient property. However, we might think this risks their making poor choices, since they do not understand veterinary services. So, we might think we should help them decide how to spend their money to best help their animals. We might also think charging fees is an acceptable part of a voluntary exchange, and that failing to collect them would unfairly deprive other animals and paying clients. However, we might think we should collect such fees only when the clients have undertaken to pay, i.e. when they were given accurate estimates or otherwise offered payment.

*

We might think we should set fees to ensure we get equivalent remuneration to comparable professionals (e.g. dentists). However, this might make our fees uncompetitive. We might think fees should be determined by market forces, with each practice setting whatever prices will maximize profit, since such competition will improve efficiencies and customer service. However, we might think this risks lower care quality standards (e.g. to compete on price) or fewer animals being treated (e.g. setting prices to capture richer owners) and damage to common resources (e.g. pollution), especially as veterinary markets do not function perfectly (e.g. clients cannot really judge the quality of veterinary services) and market forces ignore the interests of non-transacting parties. Alternatively, we might think we should set fees, and standards, at levels that lead to the best overall outcomes for animals (across number and standards), including ensuring we remain sustainable.

We might also think we should avoid making insurance claims that are above what we would otherwise charge paying clients (especially if this is illegal too). We might think doing so could disproportionately increase the premiums for clients, and potentially thereby reduce the uptake of insurance, which would mean other animals do not get beneficial treatment. We might also think it unfair to effectively charge insured clients more than uninsured ones (since the former still collectively pay for their animals' treatment, albeit through the insurance scheme). Alternatively, we might think that, although we should not increase charges for insured clients, we might sometimes choose to discount fees for some clients while charging insured clients the 'correct', undiscounted fees.

*

We might think we have no responsibility to care for animals when there is no owner to pay, but this leaves such animals uncared for, unless other people (who are also not their owners) pay us, and it is unfair if some vets help and others do not – plus caring for such cases, even if not a duty, is still a morally nice thing to do. We might think we should all help some unowned animal for free, as part of a general veterinary responsibility that comes with the privilege of being able to practise (or because doing so is better for our reputational and staff morale, or is a regulatory obligation), but we think we should prioritize which cases and treatments, to avoid overly affecting our practice's sustainability.

Some treatment might be funded by charity donors (e.g. animals for rehoming). We might decide we should at least provide such treatment at cost, rather than gaining a profit from charity cases. We might think charities are not morally obliged to pay for treatment, and we cannot morally insist they fund treatment for animals brought into our practices (unless we are donors and think they are misusing charitable funds on less valuable work). We might also consider there should be a pragmatic limit on what charities should be expected to provide, as helping one animal uses resources that could be used on other, needier animals, or excessive subsidized treatment might affect the charity's financial sustainability.

Ideas

 Confusing origins and ends
Sometimes we confuse our cause with what we should do. For example, we might think we should allow meat-eating or cut-throat commercial competition because of how we evolved, or seek profit because our practices were funded by speculation.

 Race to the bottom
Competition can mean that standards get lowered to the lowest common denominator, to the disadvantage of all.

 Cutting costs
Charging clients can feel embarrassing or stressful, and is an aspect of veterinary care that many vets dislike. Charging can also elicit emotions from clients such as indignation, guilt and frustration. Not helping animals for ostensibly monetary reasons can be stressful and make us feel personally responsible. We should ensure our prices are fair, but then stick by them as a legitimate part of private practice.

 We're worth more than gold
We sometimes feel as if lower fees or salaries imply we are less valuable (e.g. than dentists). We should value fairness, while also recognizing that money does not accurately measure real value.

Applications

- Would our level of care (for a given price point) be better than that of an 'average' vet?
- If our standards risk us losing clients, should we alter our business model or client base?

- (How much) would subsidizing this work really reduce our ability to help others in future (e.g. affecting practice sustainability), create an excessive demand, or be an unfair sacrifice for us? (How) should we limit the cumulative effect of providing subsidized treatment on practice sustainability? Can we get other clients' permission to cross-subsidize? Should we set up a designated fund to help needy animals?
- Is there an owner who is responsible for funding their animal's treatment? Can they pay (or do they simply choose not to)?
- Could we use the subsidy to help more or needier animals (e.g. who need more basic treatment, or unowned strays, wildlife and animals in rehoming charities)?
- Would providing pro bono treatment benefit the practice indirectly (e.g. for morale or public relations)? Is the upfront financial variable cost actually worth it?
- Is a certain amount of pro bono care an expected (or professionally obliged) part of being a vet (e.g. providing first aid to unowned animals)?
- Would it cost us less just to treat this animal for free than we would charge a charity (considering taxes, administrative costs, etc.)?

Veterinary Research and Education   **11**

11.1   Deciding whether to undertake animal research

As researchers, we use animals or animal tissues. As official vets in research institutes, we might have specific responsibilities to safeguard animal welfare. As clinicians outside laboratories, we apply findings from physiological, pathological and pharmacological studies to our patients, and prescribe drugs that were developed and tested using animals or tissues. As policy-makers, we use animal welfare science and epidemiological data in writing policies and providing opinion and education. And all of us use cognitive, anatomical and ethological data in establishing our ethical views (e.g. based on evidence of which species seem able to suffer). As veterinary professionals, we also affect how research is viewed, conducted and used in society.

Let us imagine that, for whatever drugs (or data) we were using in our earlier case, we are now working as a Principle Investigator developing those drugs and efficacy data. How should we consider the animals used in research and drug testing?

- When should we consider research to be acceptable? On what should it depend? Can we even make such assessments?
- What limits should we place on how any individual is treated? How (should) we minimize harmful impacts?
- (How) should we use data or products that research has developed?

We might believe that animal research is always morally right simply because data *can* be useful, but this view ignores the actual likelihood of such effects for a given study, and other unintended outcomes (e.g. any suffering of the animals involved). Conversely, we might believe laboratory animal research is never acceptable simply because some project methodologies involve harming animals (e.g. by creating disease models, keeping animals in laboratory conditions, using them in experimental procedures, 'sacrificing' them for tissue samples, or destroying surplus bred stock), but this ignores the potential benefits. Alternatively, we might believe some veterinary research is ethically justified and some is not.

We might think we should only do research if it leads to the best overall outcomes for all animals and people considered 'as one patient'. This would justify projects when the data help patients more than the methods harm subjects *and* when this net advantage is greater than alternative approaches for gathering data (e.g. *in vitro* or observational methods; clinical trials using pre-existing pathologies; use of tissues rather than live animals; study designs using fewer animals or less suffering) *and* better than avoiding the issue in the first place (e.g. public health efforts to prevent the conditions occurring, for example through human lifestyle changes). It would mean projects should not occur when the risks outweigh the benefits overall or where there are any other better methods for gaining such data or otherwise avoiding the problems we are trying to solve.

However, we might question our ability to make such assessments accurately, especially given the unpredictability of research and its translation into practice (particularly for 'basic' scientific research), the incalculability of long-term effects (e.g. how many patients it might help) and what else needs to happen for those benefits (e.g. studies might help nobody if the data are unpublished). We might also worry that our assessments might risk significant bias (given that project evaluators are often scientists and always humans). We might think we should therefore 'err on the side of caution' in protecting research animals, or even that we cannot confidently justify animal research based on potential benefits (logically, it would be retrograde reasoning to conclude we can perform animal research without such assessments). Nonetheless, we might think we can confidently assess research by comparison with other methods.

We might also believe that, notwithstanding the overall calculation of outcomes, we should set limits on our treatment of any one individual. We might think no animal should be caused severe suffering or a life-worse-than-death (e.g. by providing opportunities for pleasant social and environmental interactions to outweigh any suffering involved in the experimental procedures). We might think no subjects should be used in harmful experiments unless they have given consent (e.g. adult human clinical trials). Indeed, we might hold this view even if we do not believe that animals can give meaningful consent, and then conclude that all harmful animal research is unethical (or, at least, that animals should be given additional protection, like children in medical trials). Logically, the view that animals should not have such protection because they cannot consent is muddled (it is actually the absence of a reason to ignore a reason not to harm them).

*

Whatever our view on the uses of animals as experimental subjects, when it does occur, we might think we should minimize any harms to the animals used and maximize the benefits to patients. We might think all experimental treatments or scientific interventions should use the fewest animals needed for meaningful data (including not repeating studies unnecessarily);

optimal breeding, husbandry, palliation and clinical care; careful monitoring (e.g. for side-effects) and specified contingency plans in case of any difficulties. We might think we should conduct some studies (even when we think they should not occur) so that they occur in a better-regulated environment, so as to minimize harms. However, this argument would logically only apply to mutually exclusive comparisons (e.g. where funders could fund research in our country or institution rather than in another) and not where research might get conducted in both environments or institutions.

We might also think research should maximize the benefits, through optimal study design, publishing and valorization. Logically, such mitigations do not mean a study is morally acceptable (although their absence may make studies unacceptable insofar as they lead to suboptimal outcomes). This might sometimes suggest the conducting of additional research (e.g. for clinical trials or to satisfy regulated testing requirements). We might oppose these requirements as unnecessary hurdles that delay treatments getting to patients, but we might still think we should satisfy them in order to deliver on the benefits of the research. However, this suggests we should consider these harms in our initial assessment of whether research will be beneficial overall.

*

As clinicians and policy-makers, even if we consider some research is wrong, we still have a separate question of whether to use the data or products it has generated. We might argue that the data and products are produced now, so there is no harm in using them (while perhaps campaigning for future improvements), unless we expect that market forces would mean our usage would lead to more being developed in future. We might think our responsibility is to our patients, and it is the researchers' responsibility to avoid suffering in laboratory animals. Alternatively, we might believe we should avoid using such products that we know involve harms. (We might compare similar arguments that it is fine for consumers to eat animals kept on poor farms, because they are already dead or because poor on-farm welfare is only the farmers' responsibility.)

Ideas

 Exaggerating potential benefits
A project's harms might be reliably predictable, but the benefits can be unpredictable, unspecific or optimistic (e.g. 'cure cancer'). We should instead be realistic about the real likelihood of that study leading to benefits that would not otherwise have been gained.

 Confusing the whole with the part
Sometimes we confuse the value of something with the value of part of it (e.g. that animal research is valuable because some specific studies have been).

continued

Doctor Moreau
Laboratory research evokes very visceral horror and pity in many people, perhaps prompted by images of (sometimes historic) research. We should engage objectively, trying to improve the standards of laboratory animal care (and research).

 Vets hurting animals?
We might feel conflicted between our general motivation to help animals and our specific role causing laboratory animal suffering (and tempted to assume animal research is acceptable because we use it). Whatever our views on animal research, we should engage with the issue and not ignore it.

Applications

As researchers:

- Is *this* research genuinely expected to benefit some animals? Is this benefit genuinely likely? What else would need to happen for that benefit to occur, and is that expected to happen?
- Can we ensure research animals involved have enjoyable lives overall?
- Could the same benefits be achieved using less harmful methods?
- How can we maximize the beneficial impact of the data?

As clinicians:

- By using data or veterinary products, are we morally partly responsible for encouraging the suffering of future laboratory animals?
- Do the harms to the laboratory animals outweigh the benefits to our patients?
- (How) can we encourage laboratory standards to improve?

11.2 Selecting which animals to study

As well as research in the laboratory, we can obtain data from clinical patients by performing additional tests or interventions.

Returning to our clinical case, let us imagine that a medical colleague suggests a completely novel, experimental procedure for our patient. Should we try it?

- Which species should we use in our research? Is it better to conduct research on the same species as the expected patients?
- Should we change our treatment of our patients in order to gain data, or should we only do what helps our patients clinically?
- (How) should we then use data that we have obtained as a side-effect of veterinary treatment?
- What permission to use patients or data should we get from their owners?

We might think it is preferable to use members of particular species in research (e.g. mice or zebrafish) to benefit another species as patients (e.g. cats or dogs), because of biological and psychological differences (e.g. cognitive capacities), practicalities (e.g. cost to keep and speed of reproduction) or humans sentiments (e.g. dogs are generally more beloved than rats). However, we might worry that such differentiation represents irrelevant, unjustified emotional predispositions (as most vertebrate species seem equally able to suffer), or applies only because of circular ethical assumptions (e.g. they are cheap to keep because we house them sub-optimally). Alternatively, we might think research is more acceptable when the same species benefit, to avoid bias and facilitate extrapolation. However, we might think species is irrelevant when studies still harm some *individuals* to benefit others (from the laboratory animal's view – I would not think experimenting on me more acceptable if it helped humans).

∗

We might think we should gain data from actual patients (e.g. through clinical trials) because the data are more validly applicable to other cases, and because it avoids creating disease models or keeping animals in laboratory conditions. However, we might worry that field-based research has insufficient regulatory controls to protect patients (compared with laboratory research). Conversely, we might think that, as clinicians, we should select treatments based on what we think best for our patients. This would suggest we should not use patients for research unless we expect it will not harm them (e.g. purely observational or epidemiological studies), where the risks and benefits balance out, or when it will benefit them as the best – if uncertain – bet (e.g. if the unknown risks are worth taking). More widely, we might also think we should avoid the potential (perceived) conflicts between our roles as clinicians and as researchers that researching on our patients creates.

∗

Even if we think our clinical decisions should ignore the opportunities to produce publishable data, we might still think we should subsequently decide to publish any data that we do obtain by providing treatments, as we are not harming our patients by merely sharing their data. However, we might also worry that an intention to publish data could bias our clinical decisions towards procedures that are not in our patients' interests (e.g. providing experimental therapies, taking additional samples or delaying euthanasia).

∗

As well as considering our patients, we might also think we have a duty to protect our clients' animal property and data, and to avoid owners being disinclined to present their animals to vets in future because they are worried that we might experiment on them or share their data. We might therefore think we should experiment on owners' animals (including cadavers), or use clients' data, only when they have given us permission. However, we should avoid confusing this view with the reverse viewpoint that owner permission means experimentation is right, or even acceptable. We are still responsible for any harms we cause in trying novel therapies on our patients, even if owners have no objections or suggest we try a novel approach on their animal (based on their internet surfing).

Ideas

 Justifying justifications
Since we can come up with some justifications for almost any decision, we might try to evaluate whether each justification is itself justified (and so on). Generating data is a justification only if that justification is justified in the specific case.

 Corrupting biases
Sometimes we have some self-interest in decisions that we might make better if we were personally unaffected (e.g. gaining research publications from our clinical work), or have loyalty to someone where we should be impartial (e.g. assessing colleagues for misdemeanours). Such conflicts might be actual, potential and perceived.

 Me and my dog
We may dislike the suffering of pets (and humans) above that of other animals, because we can imagine our beloved companions (or selves) in that situation. However, we might think we should ignore this emotional bias when animals have comparable abilities to suffer.

Applications

In ascertaining whether laboratory research is better, we might ask ourselves:

- Is there any reason to be less concerned with equivalent suffering in different species (e.g. humans against other vertebrates)? Are there biases we should ignore?
- Is there any reason to believe that the overall suffering would be less in one species than another (given the methods and oversight likely to be actually used for each species)?
- For a novel therapy, do we genuinely believe it is the 'best bet' for a particular patient, based on our clinical judgement and any available data?
- Might we have any biases in our clinical decision-making (or research methodology) due to our desire for data (or concern for our patients)?
- What are the regulatory frameworks that cover clinical and laboratory research?

11.3   Deciding what to learn and teach

What makes us vets is a combination of our knowledge, our viewpoints, our judgement, our skills and our behaviour. We are required to learn and teach significant amounts of knowledge and develop considerable skills. We might also engage in specific learning or teaching activities in relation to ethics as a subject.

Let us imagine we have a student with us for our case (e.g. that we are working in a teaching hospital). Let us also anticipate that we are considering our own continuous professional development. What should be our learning objectives?

- Is ethics a topic we should study and teach? Should we teach it as its own subject or within other courses?
- When teaching, should we reveal, propose or impose our moral views, or hide them?
- Should we tell students what to do, or help them develop their own views?

We might think we should formally teach veterinary ethics to students and colleagues (and clients) and learn about it ourselves. We might think ethical competencies are helpful in improving our decision-making, communication, teamwork and personal wellbeing. Alternatively, we might think we should teach practical ethical skills and help students learn to make robust decisions, to recognize, understand and compare others' viewpoints and to articulate, reflect, critique and justify their own. We might hope to improve their consistency, confidence and resilience under pressure and reduce moral distress.

We might think we should teach ethics as its own non-clinical subject, since it is a separate field of independent enquiry. Alternatively, we might think we should include ethical skills training in clinical subjects (and other subjects and skills such as business, communication and professionalism), alongside the relevant scientific facts (e.g. anatomy, pharmacology and reported outcomes). We might therefore want clinical specialists to discuss and explain their decision-making in current cases in real time. We might also think specialist and clinical teaching staff would want, and need, to have developed their ethics skills in order to discuss ethical aspects of the cases competently and confidently (or sometimes teach alongside specialist ethicists).

*

We might think we should avoid imposing our own views on students, recognizing cultural differences, humbly accepting that our views might not be right, and trying not to stifle students' personal reflection or discussion. However, we should recognize that the argument that we should be ethically neutral is itself an ethical viewpoint (so saying we should be ethically neutral is, paradoxically, not ethically neutral), that we probably cannot completely hide our views, that students may want guidance and role models, and that trying to withhold our views might make students wrongly assume we hold different moral views (e.g. if we do not comment on poor animal care, students might think we endorse it) or that we consider taking a moral view is of no importance.

Alternatively, we might think we should be open about our moral views when teaching ethics or any other subject. We might believe that this will improve consistency and compliance, that we should promote our own ethical views (because, by definition, we think they should be followed), or that it will help students to develop their views. Specifically, we might think we should express our views and explain our underlying factual beliefs, moral beliefs, assumptions and reasoning (and its gaps). We might also think we should role-model not only what we consider to be moral behaviour (by being honest, compassionate, etc.) but also skilful ethical thinking (by being open, informed, consistent, patient, courteous and humble).

*

We might think students should be told what they should morally do, and need to have their views formed or corrected through instructions (e.g. practice or professional rules) or persuasion. Alternatively, we might also think we should help them develop their views (and let them challenge ours), building up from their fundamental views to well-grounded conclusions, even when we disagree with them, by coaching them to analyse their own views. Indeed, we might think we should actively encourage students to challenge our views, to help them understand their bases and logic, form their own views, and practise critiquing ethical viewpoints (without them publicly criticizing one another's). We might also expect this could improve our ethical views and skills.

Ideas

 Hidden curriculum
We teach students in implicit and unintentional ways through our statements and behaviour, sometimes reinforcing a wider culture (which might involve a lack of compassion, integrity or ethical reflection). We should instead ensure we are modelling mature ways of thinking, and acting.

 Being too close
We often look to specialists for views on topics in their area, as they should best know the facts. However, specialists are not normally specialists in ethics. Some may even seem 'too close' to controversial practices or have a conflict of interests of loyalty. Instead, we should rely on experts' factual information, but draw on other ethical views.

continued

People, not positions

Sometimes we are predisposed to accept or reject views because of who holds them (e.g. automatically rejecting views held by vegetarians). We should instead listen to everyone and then form our own views based on the arguments. (We should also not assume we know best simply because we are vets – or ethicists.)

 Open to openness

Learning and teaching ethics can make us feel awkward, embarrassed and vulnerable, and we may feel judgemental or sanctimonious in defending or proselytizing our views (or we may not be that confident in them). We should manage conversations so that they feel like a collaborative exploration of difficult issues.

Applications

- Is our experience improving our ethical thinking and behaviour enough? Would more reflection, discussions or formal learning help?
- Do we have the required factual knowledge, experience and ethics skills to be credible and effective? If not, how can we obtain them? How can we help other clinicians develop their ethical skills and confidence, and facilitate skilled ethics teaching (and so provide positive role models)? Can we highlight ethical aspects of real and hypothetical teaching cases?
- What is behind our students' views? Would it be effective – and not too risky of harm – to challenge students' views in class? Do we need to guide discussions, so that everyone challenges their habitual reasoning patterns and preconceived views and maintains respect for each other?

Veterinary Policy-making and Enforcement

12

12.1 Making policies

Most practices and countries have policies that are intended to prescribe or proscribe certain behaviours (e.g. laws, veterinary professional rules and practise Standard Operating Procedures). These might cover, for example, animal care and movement, property, financial transfers, employment and personal data. In many cases, vets are involved in making those policies, particularly animal laws and veterinary professional rules.

Reflecting back on our case when the owner did not want what was best for their animal, let us now imagine that we are involved in setting the practice or regulatory policy on how vets should make decisions (e.g. whether they can provide treatment without an owner's permission). How should we go about making such rules?

- What moral force do our professional policies have? Are they mere agreements, do they describe what is ethical, or can they create moral responsibilities?
- How prescriptive should our policies be? How much should our policies restrict what individuals do? How detailed should we make them?
- How do we ensure our policies can all be followed, in particular that they do not conflict with one another? Should we write them negatively (i.e. 'Do not') or positively (i.e. 'You must')?

We might think of our professional codes of conduct, and practice policies, simply as 'rules of etiquette' or 'club rules'. This suggests they are not moral per se (if hopefully not immoral), but just mutually agreed criteria for membership. However, this view of veterinary policies is unlikely to satisfy those of us who think that being part of a profession implies adhering to particular moral values. It also suggests we should ignore policies when we think there are better options.

We could go further and think policies should simply describe or articulate what we (as regulators or the public) consider moral. Descriptive policies might make our decision-making easier by providing 'rules of thumb' that lead to the best outcomes on average. They might also help us to avoid pressure from others (e.g. if we inform them that what they are requesting is illegal). Furthermore, they might motivate anyone who does not share the described ethical view, if following them is linked to their personal outcomes, either to avoid personal disadvantages (e.g. disciplinary sanctions or fines) or forfeiting advantages (e.g. losing accreditation status). Descriptive policies might also link our moral duties to our privilege to practise veterinary medicine (e.g. looking after strays or providing first aid as professional duties).

We may go even further and think that professional policies add to our moral responsibilities, when they improve our coordination and reliability, helping us make better decisions (Fig. 12.1). For example, if we know other vets will provide analgesia and charge adequately, we can be more confident in refusing to provide cheaper surgery without painkillers. By the same token, we can collectively drive societal change if owners know they cannot get mitigations for bad practice from any vet. Being reliable can also increase public confidence to use our services, so that we can help animals.

*

Whatever the basis of our policies, we might think we should ensure that our policies never prevent vets from doing what is best in the particular case (e.g. practice protocols that direct overtreatment or forbid pro bono care for strays). So we might think each policy should allow ethical 'tolerance intervals' that permit some flexibility (e.g. as rules-of-thumb guides). However, we

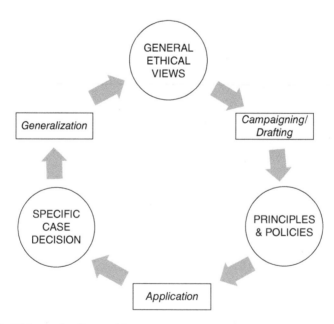

Fig. 12.1. Ethics and policy-making.

might think that sometimes policies should even limit morally acceptable options, where the benefits of improving coordination outweigh the benefits of allowing autonomy in any single case. For example, a rule that forbids all vets from a mutilation may stop vets using this to help some patients, but force owners to provide enrichment.

We might think brevity and simplicity are better. However, this approach might oversimplify complicated issues or overgeneralize to the point of being less clear for vets deciding what they should do. Alternatively, we might think that we should specify, in detail, exactly what is required. For example, rather than a blanket rule requiring owner consent for treatment, we might specify multiple rules that: (i) owners are kept informed and engaged in decision-making, and their views are considered in determining what is best for their animals; (ii) clients' money is not spent without their knowledge; and (iii) vets do not perform any unethical treatment (i.e. they need to be confident to override an owner's disinclination). However, we might worry that this would lead to long, unmemorable lists of rules.

٭

We should also ensure our policies do not conflict with one another, to avoid causing moral distress or making complete compliance impossible (e.g. that vets should prioritize animal welfare *and* gain consent for all procedures). We could try to make our policies with sufficient detail to avoid any conflicts (e.g. listing all exceptions to each rule), but they would need to be extremely complicated given the complexity of practice. We could have only one positive rule, and the rest negative (as we can then always comply by doing nothing), but this suggests veterinary work has only one goal. We could allow each vet to decide between conflicting policies, but this might seem unhelpful and confusing for vets worrying about non-compliance, and undermines all the policies as effectively ignorable. We could create an additional policy on how to resolve conflicts, but then we might think it just makes more sense to use that as our single policy.

In practice, we might use a combination of policies. For example, we could create a single basic positive aim as a 'bedrock' (e.g. 'Do what best helps your patients') with some constraining prohibitions as 'red lines' (e.g. 'Do not lie to clients'), with which each vet must comply (or not be a vet). We might then add layers of guidance that suggest a particular choice, but allow individuals to exercise their judgement, taking into account each situation and their own personal ethics.

Ideas

 One rule for all

We may feel ethical views apply to all relevant cases. Of course, consistency is only one concern: we might think it better to be right sometimes than to be wrong always.

continued

Outcome-based rules

Some rules are based on an expectation that outcomes would be better, overall, if everyone followed them. However, this raises questions over what to do when we could cause better outcomes by breaking them – if we break the rule, we undermine it; if we follow it, we undermine its basis.

 Feeling of power

Writing policies can feel intolerant and sanctimonious (and can elicit a dangerous enjoyment of power and authority). It can be tempting to be too restrictive or too permissive (sometimes because of pressure from people who do not want their behaviour restricted). However, taking on this responsibility involves making judgements.

 Sword of Damocles

Having policies enforced can feel restrictive, patronizing and offensive, and the process can be stressful. It can feel even worse to think we might be sanctioned without being clear what we are expected to do. We should simply ensure we always do what is morally right.

 False positives

Sometimes we treat (or present) negative prohibitions as positive policies (e.g. 'Meet the needs of animal under your care' or 'Gain consent for treatment' are actually 'Do not take in animals/provide treatment without meeting their needs/consent').

Applications

- What set of policies would, if everyone followed them, lead to the best outcomes for animals?
- What degree of choice should people be allowed? Do the rules allow that, or are they too permissive or restrictive?
- If we are making factual assumptions or predictions (e.g. what would maintain public trust), on what evidence are they based?
- Do any policies conflict with one another? If so, can we alter either (or both) to avoid this conflict? If not, how should people judge which to follow, and how can we make that clear?
- Should a policy allow any exceptions (for different situations or people)? If so, can we modify its scope to make them clear? If not, how can we make it clear that it is only a rule-of-thumb?
- Is our concern about what should be done, or what should not be done? Is this a shared or unequivocal concern or one's own personal ethical view?

12.2   Enforcing policies and laws

There seems limited point in having policies unless people follow them. This might come from people simply knowing the rules exist (and deciding to comply). But sometimes, authorities might have to actively ensure that others follow them. Sometimes we have a role as enforcers (e.g. on disciplinary committees on regulatory bodies or in a practice management role). At other times, we are faced with potential wrongdoing by colleagues or clients, although we often lack information as to whether the wrongdoing has actually occurred (e.g. we have not conducted an investigation or are not experts in the policy).

Let us now imagine that we have a role in enforcing a client's or colleague's compliance with the policies we helped create. (How) should we do so?

- As regulators or bosses, (when) should we enforce policies? What about if we think the individual was actually doing what seemed morally right?
- As practitioners, (when) should we report our colleagues or clients whom we suspect of breaking policies? What about when we are uncertain that they actually have broken a rule?

We might think that we should enforce policies for various reasons (Table 12.1). Indeed, we might think we should even enforce policies that we think should be changed, so compliance is seen as always non-optimal. However, we might think such strict enforcement is unnecessary or unfair, especially when people have done what seemed right (and our policies need changing). Alternatively, we might think we should not enforce policies (Table 12.2), especially when we think someone has acted morally. However, this requires us to make additional personal moral judgements (or if those moral judgements can be codified, they should be included in the policy). Combining these, we might evaluate the overall outcomes for each enforcement option relative to other options (e.g. advice), using more severe sanctions only when other options are unsuccessful, impossible or contraindicated. However, we might think these calculations are impossibly complicated and potentially unfair.

*

If we suspect clients or colleagues of breaking the rules, we might think we should report them to the relevant authority, as part of a collective responsibility to enforce policies. However, we might be uncertain whether they have definitely broken a policy, or think that 'policing' people is not part of our role, or that reporting clients or colleagues could weaken trust in, or between, veterinary professionals, or that other methods might be better (e.g. peer–peer discussions). Conversely, we might think we should not perform such behaviour, to avoid harming others, to avoid personal harms (e.g. repercussions or damage to the practice's reputation), to show mercy, loyalty or tolerance to them. However, we might think that concern for loyalty and tolerance should not apply to ignoring immorality, and when other methods are unsuccessful, we still need to decide whether to report them.

Table 12.1. Reasons for enforcement.

Reasons	Factors that weaken this argument
Wrongdoers deserve to be punished	We think the policy is not aligned with what is morally right We are uncertain whether the person is guilty of the wrongdoing
Enforcement is needed to facilitate immediate treatment (e.g. if linked to removing animals from abusive owners or authorizing euthanasia).	When these processes can be separated (e.g. animals can be removed without prosecution)
Enforcement will prevent further non-compliance by that wrongdoer (e.g. banning them from practising or owning animals)	There are other, better methods available (e.g. advice) The risk/severity of future non-compliance is low
Enforcement will generally reduce wrongdoing by linking compliance with personal interest	The advantages of wrongdoing still outweigh the risks of non-compliance for the individual
Enforcement will help maintain public trust	The public are unconcerned, unaware or would actually sooner the individual was not punished
Enforcement will compensate victims for losses or harms	There is a better mechanism for this (e.g. civil compensation rather than sanctions)

We might think we should not report anyone unless we are sure that they 'are guilty' of non-compliance, or that reporting will lead to the best outcomes overall. However, we may lack information or knowledge to make such an assessment confidently, and consequently fail to report genuine wrongdoing that the authorities could have investigated, proved and stopped. Alternatively, we might think our role in reporting or disciplinary proceedings should be a purely procedural one, in which we raise concerns with the people best placed to investigate and judge (e.g. the police, bosses or regulators). However, this might still make us feel complicit in any subsequent outcomes, especially if we do not have faith that the regulatory rules or enforcement processes are fair and humane.

Ideas

 Dog-matic
We might think some rules should absolutely and always be followed.

 Wagging of fingers about rules of thumb
Sometimes we treat guidance as if it should always be followed. Instead, if we can think of any plausible moral exceptions to a policy, then we should be clear it is guidance rather than a strict mandate or prohibition.

continued

The invisible paw

A common idea in economics is that selfish transactions create over-all benefits for all 'traders' (e.g. in commercial trade). However, when everyone does what is best for themselves, the end result can be worse for everyone, for example through competition or when we each con-tribute to diminishing shared resources.

 Feelings of power

Enforcers can feel anger, indignation or sympathy for wrongdoers. Those being reported, investigated or sanctioned can also feel anger, betrayal and revengefulness. Those who report them may feel disloyal or suffer abuse or stigmatism from colleagues. We should try to remain objective, and ensure enforcement processes minimize harms for all involved.

 Discretion

Clients may give us histories more confidently if they do not think we will pass on their personal information unnecessarily or for personal benefit. However, this confidence may not require absolute secrecy. For example, we might think we should pass on information, when doing so would prevent an illegal or harmful activity and/or lead to the best overall outcomes.

Table 12.2 Reasons against enforcement.

Reason	Factors that weaken this argument
It is unfair to punish someone for doing what they thought morally right	If the rules and enforcement are seen as justified
Enforcement would inhibit others from doing what they think right	The rules coincide with what is right
Enforcement will discredit the rule or regulator as being impractical, immoral or draconian	The rules and enforcement can be explained
It is arrogant and intolerant for us to force others to comply with our morality	Not enforcing would tolerate intolerance or unfair harms
Enforcement would harm the individual (e.g. stress and sanctions)	These harms are deserved or minimized
Enforcement would prevent future beneficial outcomes (e.g. preventing a vet from helping animals in future)	The individual will do more harm than good
Enforcement could damage public trust (by highlighting wrongdoing)	The public are already aware

Applications

• Do we have a responsibility to ensure this policy is followed by this person? Should we actively detect wrongdoing or passively wait for evidence to be given to us?

- Could the person have complied morally and without breaking another policy?
- What would be the harms (including both the process and sanctions) to the individual? Would they deter further wrongdoing by that individual or other people? Would being merciful risk further harms to others? Would enforcing this one case actually make a difference to whatever we think the policy protects (e.g. animal welfare, consistency, public trust)?
- Is there another option that genuinely will lead to better outcomes (e.g. advice or compensation)?
- What would be a consistent response, in relation to other comparable cases?
- Could we reduce the harms involved in the process (for the victim, the wrongdoer or the reporter)?

12.3   Providing 'official' expertise

Enforcers also need to decide whether a policy has been breached. They may ask us, as vets, to help by providing facts or professional opinions. We might be asked to certify certain facts (e.g. that an animal's antibody titres are above a certain level) or provide assessments (e.g. that an animal is fit for consumption) or professional opinions (e.g. to help a criminal court decide whether an offence has been committed).

Let us imagine that the enforcement mechanism requires a professional opinion to help the enforcers make decisions (to avoid any imaginary conflicts of interest, let us assume that we are in our clinical role and not otherwise involved in creating or enforcing the policy). Should we provide our opinion?

- (When) should we provide our opinion or refuse?
- (How) should we give opinions when we are unsure?
- How should we decide what the opinion we provide should be?

We might think that we should provide our expertise because of the overall benefits to animals and society. Our information and assessments may make processes safer (e.g. legal restrictions on animal travel to prevent disease spread) or more accurate (e.g. helping courts judge if someone is criminally guilty), and these safeguards may mean beneficial activities can happen (e.g. an owner may buy a horse only if they are reassured by our 'vetting'; or importing from a third country may be permitted only if society is reassured we certified them as disease-free). Conversely, we might think we should not provide expertise when doing so would harm our patients or clients (e.g. providing an opinion that an animal is unfit to travel could reduce the owner's profit and mean the animal is then kept in unsuitable conditions or killed).

*

Sometimes we are asked to give opinions on matters where there is some degree of uncertainty that we cannot meaningfully represent. We might think we should only give opinions where we are absolutely sure that we are correct. However, such cases are rare (if even possible). Indeed, the fact that we are being asked for professional opinions suggests that there is scope for disagreement on such matters. Alternatively, we might think that we should give the opinion that we think most likely, or err on the side of one party (e.g. to protect animals or to presume innocence). However, we might think it would be better if the policies themselves did this, and our opinions should be based solely on the evidence.

*

We might think we should always give opinions that represent our actual beliefs, with integrity, honesty and impartiality. However, this might mean we have to give opinions that are unhelpful, or even harmful, for animals or our clients (while still taking payment). Alternatively, we might think we should give opinions that we think will lead to the best outcomes for individuals (e.g. whoever commissioned the opinion, or for animals affected) or overall (e.g. saying we think an animal was abused, so it can be rehomed), or what we think people deserve (e.g. saying we think an animal was not abused because we think the defendant had a valid excuse). However, this risks jeopardizing societal or judicial trust in our opinions as true, which could mean our opinions are no longer sought or considered authoritative, leading to worse outcomes in the long run.

Ideas

 Muddying the waters
Sometimes we make things seem so complex that we avoid making decisions or hinder other people getting to the truth. We should instead use our veterinary skills at dealing with complexity, to see what is important and what should be done.

 Feeling impartial
Our official work can feel judgemental or disloyal (e.g. to clients), or we might feel under pressure from others to 'bend the rules'. We might remind ourselves that our official role requires an impartial, objective approach, so being impersonal is not being inhuman.

 Objectivity
We may describe a view as being 'objective' when it is impartial (including the exclusion of any self-interest). We also sometimes use the term to mean something we can perceive through our senses and

continued

measure (e.g. bodyweight) in comparison to more 'subjective' views that require some judgement or evaluation from an individual's viewpoint. However, all views are subjective to some degree.

Applications

- Is the overall process (for which we are giving an opinion) beneficial and fair? Does that rely on our providing an objective opinion?
- What is our role and area of expertise, and what is someone else's? Are we overstepping our legitimate remit?
- Might we be biased by concerns we should ignore?
- If we are worried about potential outcomes, can we mitigate those outcomes without subverting the processes?

Veterinary campaigning **13**

13.1 Changing the world

> Many of our most difficult cases are precisely because of the ways in which society treats animals and people, due to its structure, culture and narratives. As vets, we can use our relationships and reputation to educate, enable and motivate owners, producers, consumers and citizens about animals' needs, plight and treatment. We can improve the behaviour of owners, consumers, businesses, social enterprises, charities and public bodies. We can educate, advise purchasing choices, support assurance schemes, campaign for cultural changes, and lobby for better welfare and conservation legislation, product labelling or trade laws.
>
> In our case, let us anticipate that we also have an opportunity to affect the factors that led to that case or made it difficult (e.g. the law or industry or institutional standards). Should we campaign and, if so, how?
>
> - Should we campaign on such issues (or keep our head down and focus on our clinical work or research)?
> - Should we campaign publicly or with those in the industry? How should we present complex scientific and ethical issues?
> - Should we campaign as individuals or as professions? Who should lead our collective campaigns?

We might think our veterinary moral duties relate only or primarily and directly to our patients and clients, and we should not do anything else that might distract us or risk our ability to do so (e.g. upsetting clients). This could mean we can promote beneficial changes that will particularly help our clients (e.g. reducing unfair competition on farmers), but not changes to help animals that, at least in the short term, might seem contrary to our clients' or colleagues' wishes (e.g. a farming client or laboratory head might not want us to campaign against routine antimicrobial use or invasive research procedures), potentially damaging our relationships with them and decreasing our ability to help their animals in future. However, we might recognize that some clients or colleagues might also want changes, and we should not assume everyone is opposed to change.

Alternatively, we might think we should also try to promote improvements to help animals and people, to help us avoid some moral dilemmas and make outcomes more predictable (e.g. making declawing unwanted or illegal). We might also think campaigning can morally 'offset' harms that we cause in helping our patients (e.g. when we provide Caesareans, life-saving treatments and antimicrobials that unintentionally help spread unhealthy genes, perpetuate poor management or increase competition for better farmers). This suggests that we should sometimes campaign against the very treatments that we provide, to counter their unintended effects (e.g. campaigning against irresponsible breeding).

*

In our campaign tactics, we might decide we should campaign publicly and overtly, for example to ban a practice we consider immoral. We might believe that such campaigns are the most likely ones to be effective, for example where the most likely means of change is radical societal or legislative reform. However, we might worry that such an approach would be ineffective in achieving any change. Alternatively, we might think we should work with those involved, aiming for changes 'from within'. We might believe this will be more effective, for example where commercial innovation and consumer pressure are likely to drive incremental improvement. However, we might worry that this will lead to perceived connivance.

In our campaign tactics, we might think we should present the complex facts, values and reasoning that underline our views; admit any uncertainty, predispositions and biases; be open to challenge and new information; and be clear when our views change in different cases (e.g. mutilations in different species). However, this approach might confuse people, leading to inertia or accusations of inconsistency. We might also find our messages are outcompeted by campaigners (or journalists) who are less scrupulous about cherry-picking evidence; oversimplifying issues, or using dissembling, distraction, platitudes or soundbites; and bullying tactics against opponents. Alternatively, we might think we should present our views in the ways that are most likely to be effective (e.g. dumbing down our views into soundbites). However, this approach risks oversimplifying our viewpoints, perpetuating debates on complicated matters and reducing our credibility in the long run.

*

We might think we should campaign as individuals on what we each personally believe. However, we might think this is inefficient and places unfair demands on the most concerned individuals and risks disagreements. Alternatively, we might think we should campaign collectively (e.g. through professional bodies), since we all have a stake in our shared ethical reputation, and a shared public viewpoint may be more effective in influencing others and provide 'safety in numbers' from attacks or the loss of clients. However, we might worry that efforts to find a consensus will mean we avoid trickier ethical issues, making us less insightful or progressive.

We might think campaigns to help animals or people should be initiated and led by those of us closest to them, since they have the greatest specialist expertise and specific experience. However, we might worry that this closeness could impair our impartiality and either implicitly support our clients or damage their relationship with the vets best placed to help them. Alternatively, we might think campaigns should be formulated and led by the whole profession, combining specialists' factual knowledge with non-specialists' objectivity.

Ideas

 Moral signalling

What we say or do displays particular ethical views to other people. This may be part of honest moral discussions or a side-effect of doing what we genuinely feel is right. However, we might suspect some moral displays to be merely, or primarily, efforts to improve one's reputation.

 Assuming superiority

Campaigning can make us feel morally superior (especially if we call for bans or argue in sanctimonious, priggish or sententious ways). We should (by definition) believe our ethical views are right, but we should keep in mind that we all have blind-spots, and that we probably once held different views.

 Misrepresentation of certainty

Sometimes we present facts, moral views or conclusions as if they are undeniably certain (e.g. saying 'Clearly', 'Everyone knows that …', 'At the end of the day', 'If you think about it' or 'What we've got to remember is …'). We should instead actually defend our points with evidence and logical reasoning.

 Taking flak

Campaigning can feel scary, as it involves a visibility and vulnerability to criticism (sometimes from both sides). But it can also feel bad to think we have not done enough. Sometimes we need to hold our nerve and accept the criticism as 'part of' the wrongdoing itself, as we would expect others to defend their views and behaviour.

Applications

Would societal changes help our patients overall? Would they help us avoid future moral dilemmas (e.g. by removing the reasons for particular treatments)?

- Are we at risk of perceived complicity by not addressing root causes? If we remain silent, will that imply that the status quo is acceptable or even optimal? Do we, as a profession, claim to speak out for animals and then fail to do so sufficiently?
- If changes that should happen risk harming our clients (in the short term or overall), how can we maintain our relationships with them as clinicians while achieving that change?
- Might we seem complicit or hypocritical for treating cases caused by things we say need improvement? If so, how can we explain our dilemma and reasoning?
- If someone else is spreading erroneous or unethical views, what is the fairest and nicest way to ensure they do not mislead anyone else?

13.2   Contributing to ethical debates

Thinking about our cases might also help us, as individuals and as a profession, to formulate views on some of the 'big moral issues' facing the world today, and to input into key topical debates.

Let us anticipate that, as a result of our campaigning, we are asked to contribute to wider debates on more fundamental ethical assumptions (e.g. at an academic ethics conference or as a blogger or columnist). Should we engage in such discussions, or take society's general ethical views as a given?

- Should we try to influence how society treats animals?
- Should we try to influence how society thinks about animals?
- Should we try to influence how society treats humans?
- Should we try to influence how society treats the environment?
- Should we try to influence how business behaves?
- Should we try to influence the very nature of ethical debates?

We might think we should promote profession-wide views on how animals should be treated, given our understanding of animals and our involvement in real-life decision-making, combined with our ethical skills and concepts (e.g. rather than the debate on whether farming is acceptable/wrong en masse, we can apply our concepts of lives-worse-than-death overall to propose that some animal systems should be improved or stopped, but that some others are beneficial to the animals). However, we might recognize that others' views may also be relevant (i.e. our views should not be determinative just because we are vets – especially if we are insufficiently progressive).

Indeed, we might encourage public debates to focus on the factors that are most important ethically (e.g. animal welfare rather than farm size).

*

We might also think we should use our understanding of biology and ethology to contribute to debates on the moral status of animals. Academic ethicists tend to treat all non-human species as morally equivalent 'animals' (few ethicists have had a background working in biological sciences or with animals). Our knowledge of cross-species biology and psychology should make us well placed to consider different species objectively and based on evidence, to formulate more nuanced views on which interactions with animals are acceptable.

*

Humans seem predisposed to prioritize human interests (few ethicists have had a background working with human–animal relationships, and all of them are human). We might be able to cut through polarized debates about human and animal rights, to consider ethics in a cross-species way. We might also help ethicists resolve challenging questions about how to treat the humans who are not 'normal healthy adults'. For example, we might encourage ethical discussions that do not presume that humans are exceptional or perfect.

*

Our understanding of animals and systems could also help us contribute to environmental ethics, in particular regarding the biotic components of wild ecosystems (i.e. populations of individual animals who can suffer). We might think we should use our personal exposure to the causes and effects of anthropogenic changes, our ability to balance benefits and risks of interventions, and our understanding of animals, biological systems, food production and behaviour change techniques, which we have gained from years of consultations with owners. For example, we might argue that, instead of trying to keep suffering endangered animals alive (which merely treats the symptom of the problem), we should shift our focus onto preventing environmental damage or population reductions in the first place.

*

We might also contribute to business ethics, drawing on our experiences of running private practices *and* professional veterinary work. We might share how we ensure our obligations are not only to our shareholders, but also to our customers, colleagues and animals. We might share how we ensure our commercial transactions avoid harming others, lowering standards, pollution, or depleting common resources, and how we help customers make sensible and moral consumer choices. Our practice management of clinical roles may mean we are in a rare position to identify profitable ways in which employees can be empowered to make their own ethical decisions, and how we prioritize creating value above capturing it.

*

We might even find that we are well placed to contribute to medical ethics or wider moral debates about fundamental ethical issues. Generations of philosophers have failed to come up with universally acceptable ethical views. We might be able to provide new approaches because we do not necessarily share the assumption that humans are intrinsically different (or rational); because we are both science-based and deal with unobservable matters (e.g. animal pain), we might be able to identify more sophisticated ways to combine scientific and ethical insights in order to provide genuine practical advice for real-life dilemmas.

Ideas

 Too big to swallow
Big ethical questions often feel insurmountable and interminable, but we might often find academic ethics scientifically unsophisticated or practically unhelpful. If so, that might prompt us to think how we can help improve it.

 The ultimate end
Many approaches to ethics try to identify a single concept of goodness, from which we can determine specific values (e.g. happiness, flourishing or service to God).

Applications

- Should humans, in civilized societies, treat animals in ways that harm them? Can they be changed so they are beneficial to animals? What better moral attitudes to animals can we promote?
- What would environmental ethics look like if it adequately considered the interests of animals overall, or as individuals (rather than thinking in terms of preserving species or ecosystems)?
- What personal consumer choices should we make? How can we role-model compassionate consumer behaviour, at home, at work and in public?
- How should we run our business in ways that avoid unethical transactions and obtain a fair profit? Can other businesses learn from our approach?

Epilogue

Ethics is extremely valuable. It can, and should, help us make decisions. It can feel particularly helpful when we lack self-awareness, self-reflection or self-confidence. It can prove exceptionally useful when we want to articulate our ideas, investigate those of others, and create shared views. Most importantly, it can be a way to help us avoid (or at least minimize) some of our personal predispositions and self-interests, challenge our assumptions and preconceptions, and help us see things from others' viewpoints. As a profession, better and wider discussions on ethical matters can help us develop in our ethical thinking, hopefully improving relations and reducing stress.

We must also recognize the dangers of ethics. We can misuse ethics to justify almost any position or decision. We can use it to complicate or obfuscate matters – potentially delaying or sabotaging necessary change. And we can lose confidence, faced with the complexity of the diverse and sometimes bizarre theories and concepts.

If our views about ethics have been reinforced, challenged or changed while reading this book, we should try to continue to improve through continuous reflection, self-criticism and discussion. This is hard, never-ending work. Nobody ever said being moral should be easy. (And being a vet is hard, never-ending work anyway.)

After all our investigations, analyses, discussions and planning, ethics is ultimately about love. This is not some sentimental, ignorant or biased love; it is a widespread, informed and impartial love. All our rules and viewpoints depend ultimately on such love. It can help us be patient, kind, merciful, modest, humble, honourable, selfless, peaceful, forgiving, honest, protective, trusting, hopeful and persevering.

Indeed, one might argue that the catabolic function of ethics is to clear the way of biases and misuses of ethics, in order to allow this love to flourish, while the anabolic aspect of ethics is to work out what facts help us show this love (e.g. about others, as well as ourselves). All our ethical skills of information-gathering, recognition, feeling, reflection and consultation should ultimately help us develop this love in ourselves.

Further Reading and Keywords

This book is more a taster than a textbook, and readers will hopefully want to read further.

Many original ethics books are really readable, including Jeremy Bentham's *An Introduction to the Principles of Morals and Legislation* (1789), John Stuart Mill's *Utilitarianism* (1863) and *On Liberty* (1859), and Immanuel Kant's *Groundwork of the Metaphysics of Morals* (1785). For more general ethics overviews, Simon Blackburn's *Ethics* (2001) in the OUP 'A Very Short Introduction' series provides an accessible introduction (and is almost as good as the one in the same series on *Veterinary Science*). For medical ethics, Raanan Gillon's *Philosophical Medical Ethics* (1986) is a great start, albeit out of print; a more recently revised classic and essential reading is Tom Beauchamp and James Childress's *Four Principles of Biomedical Ethics* (first published in 1985). Outside ethics, there is much to be gained from reading more on logic, choice theory, decision theory, game theory, economics, politics, psychology, philosophy and theology.

On animal ethics, Peter Singer's *Animal Liberation* (1975) is the text many non-vets say enlightened them to animal ethics, although its exposés might be unsurprising for vets nowadays. His later *Practical Ethics* has more relevant philosophical content and is shorter. Tom Regan's *The Case for Animal Rights* (1983) is interesting (if rather long), especially if read after Kant. For less analytic approaches, try *Animal Philosophy*, edited by Peter Atterton (2004). And again, David DeGrazia's *Animal Rights: A Very Short Introduction* (2002) is a good quick sketch. There are also now many veterinary ethics books. Bernard Rollin's *An Introduction to Veterinary Medical Ethics* (2006) and Siobhan Mullan and Anne Fawcett's *Veterinary Ethics: Navigating Tough Cases* (2017) sketch key ethical ideas and theories, alongside myriad case discussions. Peter Sandøe and Stine Christiansen's *Ethics of Animal Use* (2011) relates particular topics to various ethical approaches.

Atterton, P. (ed.) (2004) *Animal Philosophy*. Continuum, London.

Beauchamp, T. and Childress, J. (2019) *Four Principles of Biomedical Ethics* (8th edn). OUP, Oxford, UK and New York [1st edn 1985].

Bentham, J. (1789) *An Introduction to the Principles of Morals and Legislation* (available at http://www.earlymoderntexts.com/assets/pdfs/bentham1780.pdf) (accessed 10 July 2020).

Blackburn, S. (2001) *Ethics: A Very Short Introduction*. OUP, Oxford.

DeGrazia, D. (2002) *Animal Rights: A Very Short Introduction*. OUP, Oxford.

Gillon, R. (1986) *Philosophical Medical Ethics*. Wiley, Oxford.

Kant, I. (1785) *Groundwork of the Metaphysics of Morals* [*Grundlegung zur Metaphysik der Sitten*] [2012: CUP, Cambridge].

Mills, J.S. (1859) *On Liberty* (1st edn). John W. Parker & Son, London. [2006: Penguin Classics].

Mills, J.S. (1863) *Utilitarianism* (1st edn). Parker, Son and Bourn, London. [2008: OUP, Oxford].

Mullan, S. and Fawcett, A. (2017) *Veterinary Ethics: Navigating Tough Cases*. 5M, Sheffield, UK.

Regan, T. (1983) *The Case for Animal Rights*. University of California, Berkeley, California.

Rollin, B. (2006) *An Introduction to Veterinary Medical Ethics*. Blackwell, Ames, Iowa.

Sandøe, P. and Christiansen, S. (2011) *Ethics of Animal Use*. Blackwell, Chichester, UK.

Singer, P. (1975) *Animal Liberation. A New Ethics for our Treatment of Animals* (1st edn). Harper Collins, New York. [2015: 40th anniversary edition: Bodley Head (Penguin Random House), New York].

Singer, P. (1979) *Practical Ethics*. CUP, Cambridge.

Yeates, J. (2018) *Veterinary Science: A Very Short Introduction*. OUP, Oxford.

There are also many online resources, using search keywords such as the following.

Chapter	Keywords and phrases
Chapter 1	practical reasoning
	praxis
	normativity
	moral intuitions
	prima facie
	[Note: 'scepticism' has a technical meaning in ethics]
Part A	pro tanto
	ceteris paribus
Chapter 2.1	consequentialism
	valence
	animal welfare
	hedonic consequentialism [cf. hedonism]
	non-injury
	Ahimsa
	non-maleficence
	beneficence
	intrinsic value
	extrinsic value
	instrumental value
	utility
	signatory value
	(respect for) autonomy
	privileges
	immunities
	claims
	eudaimonia
	life worth living

continued

Chapter	Keywords and phrases
Chapter 2.2	intention
	moral accountability
	futility
	precautionality
	proportionality
Chapter 2.3	justice
	resource allocation
	rights
	rights–duty reciprocity thesis
	maximin principle
	Pareto optimality
	utilitarianism
	zero sum effects
Chapter 3.1	egoism
	altruism
	Ren
	feel-good factor
	warm glow
Chapter 3.2	virtues, virtue
	arête, arete
	De
	doctrine of the golden mean
	meta-virtues
	acts and omissions
	commission/omission
	doctrine of double/dual effect
	dual use
	cognitive bias
	[Note: 'pragmatism' has a technical meaning in philosophy]
Chapter 4	relational ethics
	ethics of care
Chapter 4.1	duty of care
	guardianship
	safeguarding
	conflation
	exploitation [using someone merely as a 'means' and not an 'end in themselves']
	complicity
	connivance
	Red Queen Effect
Chapter 4.2	legalism
	li and *yi*
Chapter 4.3	environmentalism
	deep ecology
	structural sin
	systemic sin [and 'systemic welfare issues']
	[Note: 'naturalism' has a technical meaning in ethics]

continued

Chapter	Keywords and phrases
Chapter 5.1	principles (im)partiality

Chapter 5.2	epistemology [which is similar to ethics insofar
	    as both relate to grounds for belief and action,
	    respectively]

Chapter 5.4	ought implies can affirming a disjunct denying a conjunct
Chapter 6.1	moral patients [which is one used in ethics to refer to more than medical patients] distributive question golden rule principle of generic consistency anthropocentrism human exceptionalism claim rights
Chapter 7.1	conscience moral sense superego compassion fatigue
Chapter 7.2	*Tao* *Rta* natural law *telos* teleology
Chapter 7.4	affirming the consequent denying the antecedent fallacy of the undistributed middle
Chapter 8.1	presumption circular reasoning begging the question
Chapter 8.2	casuistry *mutatis mutandis* false analogy
Chapter 8.3	prescriptivism universalizability *Grundnorm* moral pluralism reflective equilibrium converse accident particularism situation ethics mixing quantifier
Chapter 8.4	syllogism fallacy of the accident

continued

Chapter	Keywords and phrases
Chapter 8.5	cognitive dissonance
	moral trumps
	(in)commensurability
Chapter 8.6	moral dilemma
	tragic choice [and the associated feeling is often called 'moral distress']
Chapter 8.7	authenticity
	commitment
	resolution
	mauvaise foi
Chapter 9.1	moral incontinence [this is an ethical term]
	akrasia
Chapter 9.2	moral dilemmas *secundum quid*
	post-hoc justification
	escalation of commitment
Chapter 9.4	ethical perfectionism
	corrigibility
	defeasibility
Part C	thought experiments
Chapter 10.1	empathy
	diagnosis fetishism
	intervention bias
	overtreatment
Chapter 10.2	*primum non nocere*
Chapter 10.3	shared decision-making
	(valid) consent
	competence
Chapter 10.4	supererogation
	imperfect duties
	pro bono (publico)
	genetic fallacy
	caveat emptor
Chapter 11.1	cost–benefit/harm–benefit analysis
	3Rs
	fallacy of composition
	fallacy of division
Chapter 11.2	conflict of interest/loyalty
	equipoise
Chapter 11.3	appeal to authority/ad hominem
Chapter 12.1	perfect duties
	taklif
	consistency
	universality
	indirect/rule utilitarianism

continued

Chapter	Keywords and phrases
Chapter 12.2	invisible hand prisoner's dilemma tragedy of the commons confidentiality
Chapter 12.3	obfuscation certification *li* and *yi*
Chapter 13.1	rhetoric value signalling
Chapter 13.2	sustainability speciesism

Index

Note: The locators in bold and italics represents the tables and figures respectively.